Maybe Baby

Maybe Baby

28 writers tell the truth about
skepticism, infertility, baby lust,
childlessness, ambivalence, and how
they made the biggest decision of their lives

EDITED BY LORI LEIBOVICH
OF SALON.COM

FOREWORD BY ANNE LAMOTT

HarperCollins*Publishers*

An extension of this copyright page appears on page 267.

FIRST EDITION

Designed by Joy O'Meara

Library of Congress Cataloging-in-Publication Data

Maybe baby : 28 writers tell the truth about skepticism, infertility, baby lust, childlessness, ambivalence, and how they made the biggest decision of their lives / edited by Lori Leibovich; foreword by Anne Lamott.— 1st ed.

p. cm.

ISBN-10: 0-06-073781-6 (hardcover : acid-free paper)

ISBN-13: 978-0-06-073781-8

1. Parenthood—Decision making—Case studies. 2. Pregnancy—Decision making—Case studies. I. Leibovich, Lori.

HQ755.8.M386 2006

306.874—dc22

2005052686

06 07 08 09 10 ❖/RRD 10 9 8 7 6 5 4 3 2 1

CONTENTS

Maybe Baby

ANNE LAMOTT

Foreword

BY THE TIME MY CHILD WAS BORN, I had seen two ultrasound photos of him. He looked like a very nice person, perfect, helpless, sleeping. I love that in a baby. I thought about him every few minutes for eight months, I thought about holding him, smelling him, watching him grow. I talked to him, imagined the conversations we would have, and how much fun it was going to be splashing around in the ocean, and comparing notes on the mean children in the park. I lived for his arrival.

But during labor, I began to realize how hard parenthood was going to be without a partner, any money, or—what's the word?—maternal instincts. It also dawned on me that I did not actually like children all that much.

By then, of course, it was a little too late to reconsider why I wanted to have a child. Midway through the birth process, there was no way out; I couldn't just say, "Let's skip it. I think I'll go home now." This is the reason most first children get born—because by the time it's too late to back out, you have already fallen desperately, pathetically, in love with them.

I loved him intimately, sight unseen. However, by the time he lay on my chest for the first time, part of me felt as if someone had given me a Martian baby to raise, or a Martian puppy, and I had no owner's manual, no energy, no clue what to do with it.

The other part of me felt like I was holding my own soul.

Fifteen years later, this still pretty much says it.

Why did I, like many other single women, many gay men and women, many older women, and all the other not-so-obvious parents, people who used to think they could never have kids, choose to do so?

Let me begin by saying that not one part of me thinks you need to have children in order to be whole, or that there are parts of yourself that cannot be revealed any other way. Some people with children like to believe this. Having a child legitimizes them somehow, completes them, validates their psychic parking tickets. They tell pregnant women and couples and one another that those who have chosen not to breed can never know what real love is, what selflessness really means. They like to say that having a child taught them about authenticity.

This is a total crock. Many of the most shut-down, narcissistic, selfish frauds on earth have children. Many of the most evolved—the richest in spirit, and the most giving—choose not to. For parents to imply a deeper realm of living is pure arrogance.

The exact same chances for awakening, for personal restoration and connection, exist for breeders and non-breeders alike.

But some of us did have children. Here's why I did.

First, as for so many women of a certain age, my body said, Do It. Second, I assumed, even though growing up I never had a particularly strong craving to procreate beyond some *Little Women* fantasies, that someday I would be a mother. I had a couple of abortions, not for convenience but because at the time of these unplanned pregnancies I was single, broke, nomadic, and a practicing alcoholic.

But when I got pregnant at thirty-four, my assumption that I would be a mother dovetailed with the knowledge that it would be harder and harder to conceive if too many more years passed. So, I decided to do it.

I had my friends' love, and I trusted and believed in a lot of things.

My faith told me that my child and I would be covered, that God's love, as expressed in the love of my friends and family and chosen family, would provide for us, and this turned out to be true.

I trusted that even though I didn't know a thing about taking care of infants, or toddlers, kids, or teenagers, I would be provided with information on a need-to-know basis. I trusted that other parents would help me every step of the way, and that if I did not keep secrets when motherhood was going particularly badly—say, when my son decided to turn colicky, or have a temperature of 105, or during the refugee camp days of chicken pox—that there would be healing, and enough understanding and stamina to get by on. This has also proven to be true.

I also got a lot of things wrong. I thought there would be some sort of deep spiritual union between me and my child, a fleshy communion of delicious skin, of smells and textures and silences. This bond would be so rich and deep and intuitive that my lifelong quest for connectedness would finally be over. Much of this was fantasy, the longing of a lonely, scared child. But there was soft, unarmored baby skin, and that was priceless. Yet, there is so much fluid involved in having a baby—the amniotic waters, the milk, the pee, all the baptisms of infancy, what Christians call the living water—slaking our thirsts. I was unprepared for how lumpy the living water of parenthood would be, and how much grit comes with it.

I believed that being a parent would be a more glorious mechanism than it's turned out to be—that the transmission would be more reliable, less of a Rube Goldberg contraption. It's been a lot of starts, stops, lurching, flailing, and coasting, then breaking, barely in control again, gears grinding, and then something easing us forward.

When I saw and continue to see the divine spark of who my son is, I celebrate that I chose to have him. He has connected me to the child inside myself, who can still play, and live in the moment, and be silly and helpless and needy and capable of wonder. Having him also helped me to connect with my mortality, forced me to dig deeply into places within that I rarely had to confront before. What I found way down there was a kind of eternity, a capacity for—and reserves of—love and sacrifice that literally blew my mind, but also the stuff inside me that is pretty miserable. It brought me face to face with a funhouse mirror of all the grasping, cowardly, manipulative, greedy parts, too.

I remember staring at my son endlessly when he was an infant, stunned by his very existence, wondering where on earth he had come from. Now I watch him sleep on the couch sometimes when he's sick. He's fifteen now, and his feet and long legs hang over the armrest, his limbs drag on the floor, and I kind of know where he came from, only I cannot put it into words any better than I have tried here. So, I hope you will enjoy and learn from the writers in this anthology as we try to convey why we did and didn't and might have children. Who we are, and who we aren't, where we were when we first thought about having kids, or when we decided not to; where we've been since, and where we are now. This is our mosaic.

LORI LEIBOVICH

Introduction

IT ALL BEGAN WITH A LETTER.

A woman in her thirties wrote to *Salon*, begging us to publish more stories about people who had chosen not to have children. The reader, herself struggling with the question of whether to procreate, spoke of the personal cost of giving up her salary as the primary breadwinner in her family, even temporarily. "What would the return be on the investment?" she wrote. "Are there any laws that would require my children to pay for my nursing home when I am old? Are they going to be a sufficient hedge against poverty and loneliness?" This woman's letter—which was marked "not for publication"—set off a fierce debate among the *Salon* staff. Some thought the reader was crass and "emotionally crippled," and said anyone who would think about childbearing in stark financial terms shouldn't be a parent in the first place. Others defended her, saying she was being refreshingly honest about her fears. The fact that she was painstakingly examining her decision to have children, they argued, made her far more responsible than those who have kids simply because they feel they *should.*

Since the issues raised by this reader sparked such a spirited discussion in our office, we decided that the question was worth pondering in a public forum. Why have children, anyway? And should you have them if you don't feel a biological or emotional urge? If you don't, will you feel those urges later, and regret it? Does choosing not to have children mean you're selfish? Or are those people who choose to have children to fulfill themselves, or to ease loneliness, or to take care of them in old age, the really selfish ones? Are the sacrifices to your body, your finances, and your freedom worth it? And why do so many parents preach the procreation gospel to their non-parent friends? How can anyone making the decision to parent ever really grasp the magnitude of the changes, good and bad, that it will cause in their life?

We wrestled with these and other questions in a *Salon* series called "To Breed or Not to Breed" that ran in May 2003. Among the highlights, all of which appear in this book, was *Salon*'s staff writer Michelle Goldberg's essay—the confessions of a happily married twenty-seven-year-old who didn't want kids but feared she'd regret it if she didn't become a mother; Cary Tennis, *Salon*'s popular advice columnist, wrote about searching for his biological urge for children and still coming up empty, though married and settled, at fifty; Amy Benfer, a twenty-nine-year-old writer and former associate editor for *Salon*'s Mothers Who Think section, told of her controversial decision to go ahead with an unplanned pregnancy at sixteen, against the advice of those she trusted. The essays were personal, provocative, and counterintuitive.

Publishing an online magazine means instant feedback. Within minutes of posting articles, we know what our readers are thinking. In our eight years in the business, however, we have never been deluged with so much emotionally charged mail as we were after we launched "To Breed or Not to Breed." We were shocked and a little overwhelmed, as hundreds of readers flooded our in-boxes with their stories—some heartbreaking, some heartwarming, none simple—about how their choices about children shaped their lives.

Many readers thanked us profusely for recognizing and respecting the choice to remain childless. Once-reluctant couples shared their "conversion" stories about how their lives had been enriched after they became parents. One mother wrote that until she had a child, she felt invisible and underappreciated by society. A father wrote with sadness about the trouble he was having staying close to his childless friends because they had so little in common anymore. The most memorable—and tragic—letter came from a mother who wrote that she wished she had never had her son, a teenager with severe emotional problems.

What made our readers so passionate about the question of whether or not to breed? For the first time in history, parenthood isn't a given. Until a generation ago, most Americans got married and had kids, period, even if they were doing it later. But thanks to enhanced reproductive technology, greater freedom for women, the advent of the birth-control pill, and changing attitudes about what constitutes a family, people have new and often confusing choices when it comes to having children. Couples can opt out of parenthood, women can have children into their fifties, single women can procreate on their own, and gays and lesbians can start families—or not. All of the old rules about childbearing no longer apply.

We realized that the stories that appeared in the *Salon* series were just the tip of the iceberg. We wanted to dig deeper into this issue that causes so much guilt, anxiety, and strong opinions. We decided to explore in this anthology the question of whether or not to parent. While the novelty of the series came from those who were passionately against having children, the book would need to address a wider range of experiences: those who grappled with doubts but went ahead and had kids anyway; those who went to great lengths to become parents by pursuing in vitro fertilization and adoption; those whose decisions about childbearing were informed by their negative experiences as sons and daughters; those who entered into parenthood blithely only to discover that it was much more work than they bargained for; and those who were grappling with their own ambivalence about parenthood and wondering what they should do.

In a particularly timely example of life imitating art, I found out I was pregnant three weeks after I began work on the book. Suddenly, what had started as a purely professional project became a personal one, too. I knew I wanted to have children years before I was ready to have a baby. When I thought of my future, I saw children there. I remember sitting at the student union in college, puffing away on a cigarette and telling a friend that while I wasn't sure I ever wanted to be a wife, I was positive I wanted to be a mother. As I entered my thirties, this foregone conclusion took on a biological, physical dimension. I fell in love, got married, and immediately wanted to grow my new family. The ticking of the proverbial clock became louder.

It was when I actually got pregnant that I became ambivalent. Almost immediately, I was knocked sideways by anxiety—not to mention nausea, depression, and crippling fatigue. For someone who thrived on control and order, I resented the fact that my body had been hijacked—and the reality of my professional and personal life about to be upended in a mere nine months sent me spiraling. Suddenly I felt that I hadn't planned this well at all. How could I edit a book about deciding to become a parent when I felt like I was walking on such unsteady ground myself? I should have waited to have kids when my career was in a more stable place. What was my rush?

"Maybe working on this book while you're pregnant will be therapeutic," one friend suggested. "Maybe it will help you clarify why you made the choice you did." Before working on the "To Breed" series, I had never really considered *why* I wanted to be a parent, just that I would do it, some way, somehow. It was a biological, more than a rational decision.

But now, for many people, biology plays a less central role. A generation ago, if you were gay or unmarried, you didn't have a choice about parenthood. Your decision was made for you by society, by the limits of science or of your own imagination. Married women—and it was always the women—who remained childless used to be viewed as defective, or they were pitied. Today, according to Census Bureau data from June 2002, 26.7 million women aged fifteen to forty-four are childless, a record number. Other reports estimate that one in one hundred children in this country are now conceived with assisted reproductive technology. Meanwhile, thousands of single women decide to have children on their own, and there are more than two million gay households with some three million children.

Many essays in this book confront this new frontier of parenting, this wilderness of choices. I sought out writers whose work I admired, and those whose decisions about parenthood were complicated or unique. I contacted parents and non-parents, and asked them to honestly, bravely examine their choices. Laurie Abraham writes about how drinking connects her to the person she was before she had children. Joe Loya and Maud Casey look at their decision about whether or not to parent through the prism of mental illness, and ask themselves whether it's selfish to pass on "bad genes." Lionel Shriver admits that she'd rather play tennis than take care of a baby. Dani Shapiro writes about how the bad mothering she received made her fearful about becoming a mother herself.

For most of us, the decision of whether or not to parent is one of the most important we ever make. I hope that these stories will expand and deepen a dialogue that only began with the "To Breed or Not to Breed" series. My own journey into parenthood was made much richer because of this book. It was wonderful, in the first few months, to talk to some of the authors of these essays about the miraculous and tedious parts of parenthood, about the fact that my identity—the person I knew myself as—was slipping away every day and in its place was this person, this *mother;* whom I didn't know or recognize. I received an e-mail from Anne Lamott shortly after I sent her a photo of my newborn son, and she wrote, "Isn't it amazing? Isn't it hard, and kind of awful sometimes?" But there were also moments—such as a rainy April day when my son was so hysterical that I had to hold him on top of the dryer for an hour in order to calm him down—when I thought about my childless contributors and felt profound envy.

As I write this, the product of my decision sleeps in the next room. He is one year old, with a full head of red hair, sparkling blue eyes, a determined and joyful spirit, and chronic ear infections. I am thirty-four, with my first gray hairs and lines around my eyes. I am exhausted, ecstatic, guilty, and wholly in love. I am his mother. It's a word I'm not yet used to, even one year into this game, and yet here we are.

PART ONE

No Thanks, Not for Me

MICHELLE GOLDBERG

To Breed or Not to Breed

When people learn that I'm in my late twenties, happily married, and am not planning to have children, they respond in one of two ways. Most of the time, they smile patronizingly and say, "You'll change your mind." Sometimes they do me the favor of taking me seriously, in which case they warn, "You'll regret it."

I've heard this enough that I've started worrying that they might be right. After all, I'm not completely insensitive to the appeal of motherhood. In fact, I have a name chosen for the daughter I don't plan to have, and sometimes I imagine the life I could give her. Spared my mortifying suburban childhood, she'd be one of those sparkling, precocious New York City kids I've always envied. I'd take her around the world, to study languages in Europe, to see the Potala Palace and the Taj Mahal. She'd have all I wish I'd had.

My husband doesn't particularly want to be a father, but he's said that, should I ever feel the ravenous baby hunger said to descend on women in their thirties, he could be coaxed into parenthood. He's a loving and generous man, and I have no doubt he'd dote on our child if we had one. So would his wonderful family, who live within walking distance of us. They're the reason his sister, a bar owner, has more of a social life than any other young mother I've met. I think of his grandmother and late grandfather, who lived in a rambling house in rural Maine. Three generations of their adoring descendants admired them as few people admire the very old anymore, and seeing that has made me think that family can be the key to the best kind of life.

Still, the vague pleasures I sometimes associate with having children are either distant or abstract. Other women say they feel a yearning for motherhood like a physical ache. I don't know what they're talking about. The daily depredations of child-rearing, though, seem so viscerally real that my stomach tightens when I ponder them. A child, after all, can't be treated as a fantasy projection of my imagined self. He or she would be another person, with needs and desires that I would be tethered to for decades. And everything about meeting those needs fills me with horror. Not just the diapers and the shrieking, the penury and career stagnation, but the parts that maternally minded friends of mine actually look forward to: the wearying grammar-school theatrical performances. Hours spent on the playground when I'd rather be reading books or writing them. Parent-teacher conferences. Birthday parties. Ugly primary-colored plastic toys littering my home.

I can sort of see that it might be nice to *have* children, but there are a thousand things I'd rather spend my time doing than *raise* them. The daily grind of motherhood seems like a prison sentence to me. Though I have nothing but respect for the work of raising children, I don't like being around them. At least, I don't like being around most of them most of the time. Some people say I'll feel different about my own, but I'm not sure I want to take the risk. I think about Martha Gellhorn, the globe-trotting war correspondent who, when she was middle-aged, adopted a little boy in a moment of loneliness and sentimentality. At first, she was in love, like mothers are. Then she grew bored and frustrated. According to a recent biography of her, she was terrible to him, and he was the great failure of her life.

Two years ago, when I wrote a story about all this for *Salon*, I got dozens of letters in response. Most were empathetic, but a few, mostly from men, were disgusted, calling me a selfish caricature of feminism.

Those letters hurt because I suspected they might be right, that some nurturing part was missing in me. But I doubt the wisdom of bringing someone into the world just to defy that fear.

Especially when motherhood seems so hard even for those who want it more than anything. When I wrote my *Salon* story, I spoke to Rick Hanson, a California clinical psychologist and coauthor of *Mother Nurture: A Mother's Guide to Health in Body, Mind, and Intimate Relationships*. He told me that raising a child is grueling even for those who like childish things. "Most parents, men and women, say they dramatically underestimated how intensely demanding, stressful, and depleting parenthood would be," he told me.

Meanwhile, despite general social disapproval of childless women, their ranks are growing. A quarter of American women will never have children, Hanson said. The numbers are similar in other developed countries. Some of these women can't have kids, but others simply have other priorities. Three-quarters of America's childless women are physically able to be mothers, according to Hanson.

They won't always be. Fertility starts declining in your mid-thirties. Sylvia Ann Hewlett's 2002 book *Creating a Life* may have been shoddy, irritating, and smug, but it was accurate in its assessment of the dismal odds stacked against women who seek fertility treatments in their forties. After a certain age, having a baby is no longer an option.

This breeds another fear—that I'll regret my effrontery in defying the whole history of the human race. Are deliberately childless women setting themselves up for a lifetime of barren desolation?

According to the experts I spoke to, the happy answer is no.

When I started writing my story for *Salon*, I would have described myself as ambivalent about childbearing. Yet when experts told me I was unlikely to suffer debilitating psychological fallout if I spared myself motherhood, I felt enormous relief, as if I'd been let off the hook. They said that people who *choose* not to have children (as opposed to those who desperately want to have children but can't) tend to have better marriages, better finances, less stress, and are no more likely to be unhappy in old age than parents. Most people, and especially most women, have a physiological yearning to reproduce, whatever the costs, and are glad they did. Yet maybe being born free of that desire is a gift.

"Some women really do love mothering," Madelyn Cain, author of *The Childless Revolution: What It Means to Be Childless Today,* told me. "I happen to be one of them. I love being a mother. It's the greatest joy of my life, but what makes me happy and brings me fulfillment doesn't necessarily make everyone else happy."

The notion that different people have different desires shouldn't be a difficult one, but when it comes to motherhood, many people can't get their heads around it. Even Cain had trouble at first. She began *The Childless Revolution* in part because she was angered by the dismissive way her childless friends were treated, and because she was struck by the newfound social acceptance she experienced when she had her first baby at thirty-nine. Yet part of her still believed that "deep down every woman wanted to be a mother," a misconception undone by the more than one hundred interviews she did for her book.

What she discovered was that choice, not motherhood, is the real key to happiness. Cain divides the women in her book into three groups—those who affirmatively decide not to have children, those who can't have children, and those for whom circumstances never align to make motherhood happen. Citing her own interviews as well as books like Elaine Campbell's *Childless Marriage: An Exploratory Study of Couples Who Do Not Want Children*, Elaine Tyler May's *Barren in the Promised Land: Childless Americans and the Pursuit of Happiness*, and Marion Faux's *Childless by Choice: Choosing Childlessness in the Eighties*, Cain said, "The ones who decide they don't want children, they don't regret it."

Cain came to believe that lack of interest in childbearing might be biological, like being gay. "Researchers have found that within mice, there is a gene, the Mest gene. When it was in place in mice, and the mouse gave birth, it was a nurturing mother. When the mouse was Mest-deficient, it was a non-nurturing mouse. I think down the line we're going to discover that just as homosexuality is something that's physical, the same thing will be discovered about women. Why do some women melt at the sight of babies while other women are indifferent? It would seem to me it's something innate."

That's why Cain said women who don't want kids should ignore the well-meaning advice they're often bombarded with. "Don't second-guess yourself," she says. "Trust your instincts."

That might seem obvious, but the strange thing about being a woman without much interest in mothering is that many people you love and admire will tell you *not* to trust your instincts. Motherhood, they say, is, for all its struggles, an experience of such ineffable joy that those who've done it can't imagine life without it. Motherhood evangelists have a store of conversion stories. Either they, or someone they know intimately, had once been like me, cherishing their independence and impatient with children. But when bathed by the blissful hormones that accompany procreation, they saw the light and now their lives are richer and more meaningful than they ever thought possible. They say those who haven't parented can't even begin to comprehend its radiant satisfactions. And, of course, they're right—we can't.

That's what makes the decision to forgo it so hard. There are few experiences in life that come more highly extolled than parenting, so how can you ever know if you're making a mistake by rejecting it? It's fairly easy to find stories of those who regret *not* having children, but it's difficult to find a mother who will say she wishes she'd made a different choice.

"It's very rare for a woman who has children to regret having children," Hanson told me. "You will find women who say, on the one hand, 'I love my children, they're profoundly fulfilling, and I can't imagine not having had them.' But on the other hand, they'll say, 'Boy, this is really stressful. This has really strained my marriage. My health has never been up to par since I had my first or second child. I really regret the impact of having kids on my career. Having children has made me financially dependent, and really limited my options for making money.' I hear them say all those things, but you rarely hear moms actually saying, bottom line, I should never have done it."

That doesn't necessarily mean they never feel that way, though. Among the many e-mails I received in response to my *Salon* article were a few from women who said they wished they'd never had kids, but that they couldn't admit that wish to anyone. In 1975, many women told Ann Landers the same thing. A woman wrote to her with qualms much like mine—she and her husband were torn about child-bearing and asked, "Were the rewards enough to make up for the grief?" Landers put the question to her readers, asking, "If you had to do it over again, would you have children?" Astonishingly, 70 percent of her respondents said no.

Just a few weeks ago, I was having dinner at a friend's house, trying to carry on a conversation between the demands of her toddler and the cries of her baby girl. When her first child was born, she'd tried to work, but soon found she was paying nearly as much for day care as she was earning. Now, despite her intellectual ambitions and first-rate education, she's a full-time mother, a role she seems half-amazed to be playing. I talked to my friend about this essay, and about the fact that, for all the sacrifices involved, every mother seems to recommend motherhood.

"I don't recommend it," she said.

She was speaking in a harried moment. At other times, she's told me that motherhood has brought her more happiness and satisfaction than she'd ever felt before. There's no way for me, an outsider, to know which of those sentiments predominates in her hectic life, and which is more enduring.

Overall, though, Hanson told me that mothers generally aren't more satisfied with their lives than childless women are. For all the truth about the innate physiological rewards of mothering, he said, "The happy people are the ones who wanted kids and had them or didn't want kids and didn't have them."

This is true even in old age, a time when many assume the child-less will suffer alone while their peers are comforted by grandchil-dren. A few years ago, Tanya Koropeckyj-Cox, a sociology professor at the University of Florida who researches aging, completed a study based on surveys of 3,800 men and women between the ages of fifty and eighty-four. "For years we have heard warnings that if you don't have children, you will regret it later," she said in a press release. "But beliefs about childlessness leading to a lonely old age are simply not supported by our study." In a previous report published in 1998, Koropeckyj-Cox concluded that there are "no significant differences in loneliness and depression between parents and childless adults."

Besides, what some parents gain in intimacy with their children, they lose in intimacy with their partners. "Research has shown that on the average the greatest challenge to a couple is becoming parents," Hanson said. Many marriages hold together for a few years when the child is young, but they've been strained beyond repair by everything that comes from having kids, and the couple divorces, maybe by the time the kid reaches first grade. Some people think they will save their relationship by having children. It almost never happens."

He cited a study by John Gottman, a renowned expert on marriage at the University of Washington, which estimates that couples have eight times more arguments after becoming parents. Hanson said he's seen this in his life as well as his practice. "Many couples overcome all this, and having children brings them closer together. That's certainly true for my wife and myself. But during the early years—our kids are now fifteen and almost thirteen—boy, we quarreled and were emotionally distant and troubled in our marriage like we'd never been. We argued about all the issues that new parents commonly argue about—how to raise the children, who is doing more, the inevitable lack of time for an intimate relationship."

You can hear the same complaints in one of those 1975 letters sent to Ann Landers, which Cain reprinted in her book: "I am 40, and my husband is 45. We have twin children under 8 years of age. I was an attractive, fulfilled career woman before I had these kids. Now I'm an overly exhausted nervous wreck who misses her job and sees very little of her husband. He's got a 'friend,' I'm sure, and I don't blame him. Our children took all the romance out of our marriage. I'm too tired for sex, conversation or anything."

Such alienation is less likely when people don't have children. "Statistics show childless couples are happier," Cain said. "Their lives are self-directed, they have a better chance of intimacy, and they do not have the stresses, financial and emotional, of parenthood."

That, finally, is why I think I'll resist the pressure to give in and join the natural order of the world. I love my husband too much, and he's enough of a family for me. When people talk about marriage being hard work and full of compromises, I worry that some day I'll understand what they mean, but so far, I don't. That's a miracle I'm continually grateful for, and one I don't want to strain with responsibilities neither of us wants.

Still, when people say you'll change your mind, they could be right. If that happens, though, I hope it's because the desire to be a mother smacks me sideways, and not because I'm afraid of what it means that the desire isn't there.

ELINOR BURKETT

Emancipation from Propagation

BARBIE NEVER SET A HIGH-HEELED FOOT into the pastel bedroom that was my girlhood home. Nor did Pitiful Pearl, Chatty Cathy, Mrs. Beasley, Shirley Temple, or Raggedy Ann. Although my mother filled my refuge in a tony suburb of Philadelphia with all the other accoutrements of 1950s middle-class girlhood—from stitched-down pleated skirts and cardigan sweaters to circle pins and *Seventeen* magazines—she never gave me a single doll. On birthdays and Hanukkah, she wrapped up puzzles and books instead. Subconsciously, I absorbed the message: Reading and figuring things out are more important than warming bottles and changing diapers.

In case I missed that not-so-subtle instruction woven into her purchasing habits, my mother—not one to leave matters to chance—made her meaning abundantly clear in a more direct fashion: "Don't worry about getting married and having a family," she told me whenever she noticed me playing house too seriously and, later, when I seemed to be daydreaming too obsessively about some perfect future with my latest boyfriend. "Go to school, develop your profession," she counseled. "Everything else can come later."

I suspect that I would have assimilated that admonition subliminally even if she hadn't been so forthright; I knew that my mother hadn't forced her way into the Ivy League, joining the first class of female graduates from the University of Pennsylvania, looking for a husband, a house in the suburbs, and a yard filled with screaming kids. A woman of steely intellect and fierce determination, she had envisioned a life as a scientist, and I became part of the plan only after she and my father bowed to pressure from their families to create a more conventional life.

Encased in an icy shell of self-protectiveness, my mother never spoke to me of the price she paid for her dreams' denial. She died without uttering a word about whether she had been haunted by her thwarted aspirations or how she coped with an existence that was, in the truest sense, not hers. Yet even in her brittle silence, she turned those carefully guarded feelings into the most precious of gifts: She taught me to resist all pressure—social or familial—to become someone I simply was not. I grew up, then, an American anomaly: one of the few women of my generation immunized against all pressure to conceive.

Where nature and nurture intertwine, sorting out chickens from proverbial eggs is an idle conceit. I have no way of knowing whether my mother's early intervention canceled out my natural maternal instinct or whether its absence was hardwired into my genes. But no matter the cause of the psychic imprint, I can honestly report that over the past fifty-eight years, I've never once melted at the sight of an infant, nor suffered the pangs my friends call womb craving, nor wondered what my own little Rachel or David would have looked like or become. The only way I can describe my feelings toward having children is bluntly: A lifetime of utter disinterest.

As a young woman, I didn't know any of this, of course. So, yet another victim of a popular culture that used the words "female" and "mother" interchangeably, I spent my teens waiting for maternal desire to kick in, expecting it to pop up one day as a dream of pushing a baby carriage through a park, or as part of a "oh, we'll get married and make babies" fantasy built around my latest heartthrob. When such yearning eluded me as I moved through my twenties, I didn't twist myself into knots wondering what was wrong. I didn't even consider therapy for my strange ailment. I simply waited, although "wait," a term that conjures up a significant level of attention to the expectation, is perhaps too strong a word. My waiting involved none of the clock-watching anticipation I feel over, say, the arrival of the Chinese-food delivery guy or the approval of a mortgage. In truth, I might not have been aware that I was expecting something to occur if so many people around me hadn't felt a need to pester me about getting pregnant.

Public obsession with the child-rearing plans of others is a curious social tic. At least until we entered into the age of glorified public confession, Americans were politely hesitant to pry into one another's lives. Asking how much someone paid for a house or a car was considered to be the height of bad taste, as was inquiring about the progress of Aunt Jane or Uncle Sherman in rehab. Only pregnancy and plans for pregnancy were fair game for full public scrutiny. So, fellow travelers on airplanes, strangers in waiting rooms, and colleagues at work regularly inquired about my offspring.

"I don't have kids," I would respond politely.

"Oh, you poor thing."

"Don't give up. You just keep trying."

"You'll find the right man soon."

The possibility that a woman might not want children never crossed anyone's mind—including mine.

Then I reached the cusp of thirty and took stock: I had my doctorate and a good teaching job, plenty of friends, and loads of free time to travel. My life felt happy, productive, complete. I was an active feminist, for God's sake, a professor of women's history. What was I waiting for? Some social expectation that had nothing to do with my instincts or my psyche but everything to do with some arbitrary definition of womanhood that refused to become passé?

Without a grain of angst—without any drama apart from elation at the prospect of seizing full control over my body—I decided to give myself a thirtieth birthday present: sterilization.

That simple decision was more easily taken than realized in the 1970s. Even when you resisted the full measure of social expectation, society got one last crack at you, in the person of doctors trained in a

Victorian ethic about appropriate female behavior. They wouldn't hand out birth-control pills to unmarried young women. Nor would they perform abortions or tie the tubes of unmarried young women. It was do or die, conceive or abstain. I was bent on neither course.

The first physician I approached would countenance no discussion: "I don't perform tubal ligations on unmarried women." Period. I laughed, passing him off as a dinosaur from the bad old days. But the second physician—also, male—turned out to be worse. He didn't just refuse my request; he insisted on delivering a lecture on the error of my ways, complete with the old "Once you get married, you'll change your mind."

More determined than angry, I steeled myself for my meeting with a third. Loaded for bear, I was trenchant, something along the lines of: Shut up and tie my tubes. He was equally firm: No way.

"Why?" I pressed him, intent on emerging with nothing short of an appointment for surgery.

"I don't want to get sued," he answered flatly.

Flabbergasted, I demanded a more thorough explanation. He actually offered one, a heartstoppingly inventive chimera: "When you do get married, if your husband is upset that you had your tubes tied, you'll tell him that I pushed you. Then he'll sue."

I started to parse out that line of thinking, a dozen retorts flickering through my brain: Do you really believe that I'll turn into June Cleaver the minute I find my Ward? Who would believe that a feminist university professor could be coerced into sterility by her physician? What century do you live in? I bit them all back. The surgeon clearly was beyond logic.

Horrified at the reality of being infantilized—of being dismissed as some sort of eighteen-year-old ditz—I refused to succumb to fury. No way would I give my nemeses the satisfaction of my anger; it would validate their importance. Getting my way, I decided, would prove infinitely sweeter.

That afternoon, I phoned a dozen male friends and invited them all to play my husband. An old pal bit, and two weeks later, we donned matching dime-store wedding bands for our appointment with yet another gynecologist. Once in his leather-heavy office, we spoke softly, earnestly about our abusive childhoods and our fears that we would damage any offspring of our own. The doctor nodded sagely, moved by our plight. Ten days later, he rendered me blissfully, gloriously, sterile.

That sense of liberation had little to do with sexual freedom or with concern over the vagaries of birth control. Rather, I suddenly felt as thoroughly inured against social pressure as I was against pregnancy. Whenever a stranger asked if I had children or nagged me about how incomplete I would be without them, I imagined myself cutting him or her off with the ultimate conversation-stopper, a crisp: "I've been sterilized."

Yet the freer I felt, the more colleagues and acquaintances acted out their horror and disbelief, their brows shaded with questions, usually unspoken: Was she abused as a child? Molested? Unloved? What sort of pathology would lead a woman to renounce children? For the most part, men stayed silent. But few women felt so constrained.

"It's reversible, isn't it?" a female colleague, the mother of two, asked abruptly upon walking into my office unbidden. "Because you'll change your mind when your biological clock starts ticking!"

It wasn't and I wouldn't, I patiently explained, as I'd already explained to a dozen other women. At that, she began to stammer. "But you're only thirty. That was a terrible mistake!" I resisted reminding her that she had made a major life-altering decision well before that age: She'd given birth to a son and a daughter.

When I was younger, I sloughed off these responses, laughing at such a narrow definition of womanhood. I felt no anger at the refusal to accept my inclinations as both sane and legitimate. *This, too, shall pass*, I thought, sanguine about the dissolution of old stereotypes, confident in my assumption that the confusion of womanhood and motherhood would wind up in the same historical dustbin where the belief that women can't lead nations or die for their country was being laid to rest.

But with the passage of time, my confidence has flagged. Feminism has become so ingrained in our culture that no one blinks at the sight of a female construction worker or a male nurse. Mayors and clergymen now perform weddings for same-sex couples. In every sphere, American society has come to accept personal proclivities—from religion and gender to clothing—that were taboo back when my mother was avoiding the doll aisle. Yet the assumption that women are biologically programmed to conceive refuses to crumble. The right to choose seems to have bypassed the childless.

Almost three decades after my delicious fraud in the gynecologist's office, strangers still grill me about my children or hector me about their non-existence. I'm well past the age of caring whether others find me a freak or assume that I'm a child-hater. I'm too busy savoring my mother's gift by writing books and magazine articles that have taken me light years beyond the cozy confines of my America. I'm too comfortable in the warmth of an idyllic, childless marriage, of my family, my voluntary multigenerational family. I'm too happy with a life of guiltless travel, with the freedom to fly to Afghanistan in the wake of the Taliban or to plan a year's adventure in Africa, then, to feel any personal sting.

But the more the personal wanes, the more the political waxes. These days, I seethe when I'm subjected to yet another round of "Oh, you don't know what you've missed" uttered by someone so politically correct that she would never presume that anyone was straight. And when I listen to my university students struggle with the near-universal expectation that they'll reproduce, I despair that even in the twenty-first century—the century of respect for diversity—parents have an enduring need to recruit.

I try to be tolerant, even if they aren't, reminding myself that they are simply projecting, their vehement reaction to the voluntarily and happily childless a reflection of their inability to comprehend our choices. I can empathize with that utter perplexity. Back when my friends were struggling to conceive or grieving over their inability to do so, I was equally baffled.

But occasionally, the hectoring gets too much for me, and I boil over at what ultimately should be a simple matter of respect for difference. Few of us would stoop to nagging a lesbian or a gay man into "trying out" heterosexuality, or pressuring a vegetarian into carving up a turkey. Why is trying to convince me that I've violated some law of nature or that I'm in denial about the texture of my loss any more socially acceptable?

At those moments, I'm seized with the mischievousness of my old bad-girl self and imagine turning the tables, missionizing on behalf of non-parents with the same blithe disrespect that I've endured for so many years.

"Oh, you poor thing," I'd coo sympathetically to pregnant women. "Did your mother buy you too many dolls?"

CARY TENNIS

It's Not in My Nature to Nurture

WHAT DOES IT MEAN EXACTLY, TO NOT WANT KIDS? What does it feel like when I run my mind over the contour of an absence? Where is that place in the body where most people have a desire for children? I can't describe the absence; I only know what it feels like to really want something. I only know the things that on my deathbed I might regret not doing.

The novel. The book of poetry. The songs. The playing of music. The marrying of my wife. If I had not tried to do those things, I would wonder if I had missed out. They are small things, perhaps, when weighed against having children, but they are as dear to me as my own heartbeat, my own eyesight. There was no defining moment when my wife and I decided not to have kids. In fact, we have not ruled it out completely. What is notable, though, is the absence of that storied urgency. Each time we circled around the question, poked at it, tried to see honestly what we felt, each time, and there have been many, we came up strangely empty. That might sound uncomfortably literal, but what I mean is that there does seem to be a powerful wish, a yearning, for children, that most people who have them will tell you they have felt, and we don't have it.

We don't know what it means about us, but we accept it as true, and we trust it to mean that we should not avidly pursue parenthood. Perhaps it's a little like being gay: You're just this certain way, and it doesn't feel strange to you, but it's different from the way most people are. And you might be curious to have what they have, but you're not driven to strive for it. I can say only that it feels completely normal, except when we become small-minded and start comparing, when we weigh what we've got in our hearts against what they've got in their strollers.

When I met her, my wife was on the Pill, but her doctor told her she should discontinue it, because she suffered from migraines. I came of age sexually before the advent of AIDS and had never become accustomed to using condoms. So, my wife used a diaphragm. But after several years of the habitual pause in the proceedings that their use requires, we noticed that there was more going on in that moment than simply the mechanical preparation. Something troubled us about our refusal to accept the possibilities nature offered. Our practice of prophylaxis seemed, in a word, sterile. Although we didn't crave kids, we weren't terrified of the possibility, either, and we began to feel that we were rather rigidly standing in the way of one of life's natural outcomes. We had never categorically and utterly ruled out becoming parents. And there was something else, some whiff of the mystical, in our decision to stop using birth control; it wasn't entirely rational or absolute. In a sense, it was mischievous, the way two kids will explore a vacant house, not because there's something in it that they want but because it's just there and they're curious about what it would be like to walk around inside it.

We wanted not necessarily to *try* to have a child but to be open to the idea, to stop foreclosing on this potential within us. We became willing, at that point, to have a baby if that should occur. We knew, if it should happen, that we would respond to it as humans have for ages. But we were not attracted to the notion of *trying* to have a child. It was not something that, as though running a small manufacturing concern, we wanted to produce.

Nevertheless, with a certain giddy sense of revolution, we opened ourselves to possibility. It felt virtuous. It felt like facing reality. It felt like we were in tune with some larger force.

But that excitement and sense of rightness soon faded for me. In its place came a subtle kind of dread. And into the previously carefree ritual of lovemaking crept a grave discipline of acceptance of the possible consequences. I had begun to wonder how I could summon a lifetime of daily parenting to sustain a moment's philosophical inspiration. Since deciding, in a sense, to perform without a net, we lived with its lifelong consequences, even though they were as yet only hypothetical.

And this had begun to weigh on me; at least it was weighing on me until I found myself sobbing with grief and joy at the end of the film *25th Hour*.

In *25th Hour*, a young man has pushed his luck too far and is about to go to prison. But the movie shows us what it might be like if he got a chance to start over. His father, who has come to drive him to jail, could instead drive him out of the city and just keep driving. It would mean exile and a secretive life; he would have to resist contacting anyone from his past for years to come. But it at least would be a chance at freedom. And someday in the future, his lover would join him, and they would raise a gaggle of children, and life would be beautiful.

And there I was, sobbing in the dark, because those children represented salvation, and the father's action represented mercy. It would be the ultimate act of fatherly devotion, of rescue and protection, for this father to save his son from fate and consequence, from all his sins, to put him in his car and drive him down highway after highway, past city and town, beyond the system and the harsh hand of law, to say good-bye and good luck in a tiny anonymous town where maybe his son could get a job as a bartender and nobody would know his name. I found it deeply moving that the father, traditionally the upholder of laws, helped his son disappear rather than see him suffer. I came home shaken, thinking maybe we ought to make some kids, but my wife was asleep, and the poodle was on my side of the bed.

Word has reached me that my father is unhappy that none of his four genetic children has produced an heir. He has never told me this. My wife says it's not something a parent says, that it's just something a child knows. To hear that he might have been silently hoping all this time while saying nothing is a little unnerving and a little sad. My dad always said, be independent, do your own thing. I took him at his word and put three thousand miles between us. And now that he is eighty, the terms of our pact of protection have been reversed. It is my turn to look after him. But from this distance, I cannot look after him. That makes it all the more troubling that I may have let him down by doing what he encouraged.

But absent any strong prodding from my family, I simply have not been driven to have kids. And again, I find myself asking, Why is that? Why am I not drawn to become a nuclear chieftain, king of some clapboard castle, happy monarch over a freckled brood? Why can't I picture myself as a father? Is it because the picture I have of a father is an unhappy one? Is it because of lingering resentment, a desire to refuse my own father's most secret but deepest of wishes? Is it my own wish not to repeat the strange, unaccountable bleakness of my childhood? Or am I concerned that a child of mine just might treat me as I have treated my own father, wavering in my affection, accepting his generosity with thin gratitude, abandoning him in his old age?

In a recent e-mail exchange, a regular reader of my *Salon* advice column pointed out that the reason genetic paternity matters to many men is that fathering a child represents a bid for immortality. While I don't thirst for immortality through reproduction, I do thirst for it through creative acts. Still, it's all rather silly. Once you're dead, you won't care whether a curious reader fingers a volume of your poetry or a great-great-grandchild stares at your portrait and wonders who you were.

When you think about where you come from, it's really quite amazing: Some knobby fish-eating proto-Welshman sharpening a crab spear on the pebbly shore of Cardigan Bay spies a budding weaver girl cracking open clams for her father in a stone hut's shady lee and takes her in the nearby heath. That happens a thousand times, and then it's your turn.

At any rate, it looks like the buck stops here, with me and my siblings. My father's ancient line is coming to a halt. But do I hear a cosmic voice saying "Accept the compliment and pay it forward"? No, all I hear is a little voice that says, "Finish the novel."

LIONEL SHRIVER

The Baby Stops Here

I FIRST FORESWORE MOTHERHOOD when I was about eight years old. The patronizing, she'll-soon-change-her-tune smirk that this proclamation elicited from my elders only solidified my resolve. Indeed, an obstinacy—a pigheaded commitment to doing what I'd always said I'd do if only because I'd always said it—would remain with me for life. I'm now forty-seven, and I never changed my mind.

As for what prompted my precocious aversion, from this distance I can only speculate. I may have been thankful for my mother's preparation of big "Spanish noodle" casseroles, her attendance at droning PTA meetings for Fred A. Olds Elementary, her swabbing up of my vomit when I threw up in the backseat on car trips—but no amount of gratitude would have made me yearn for the day that I would get to swab up vomit, too. As for my ambitious, restive father, he was forever relishing aloud that glorious day when he could finally have "adult conversations" with his children. While he meant us to take this impatience as a compliment, I couldn't help but reflect that it would have been more efficient by half had he conducted his "adult conversations" with adults, period. Moreover, he made little effort to disguise the fact that until that day arrived we were an annoyance. But rather than feel wounded, I think I sympathized. We *were* annoying. We were loud and sneaky and broke things. As an eight-year-old, maybe I was simply mortified by the prospect of being saddled with myself.

But what has continued to frighten me off children for all these years? Big, fat fears are often bundles of smaller ones, and any woman contemplating what never used to be a choice could rattle off the downsides of motherhood: hassle and expense (and not only of money). An acceptance that comes reluctantly to boomers of being a grown-up for keeps. The relegation of one's own ambitions so far to the back burner that they fall off the stove. A precipitous social demotion that I inferred from the chuckle of those smarmy adults who discounted my renunciations at eight: *You say you want to be a writer, but you're a girl, and all you really want to be is a mommy.*

Yet, however strenuously I might defend a woman's right to refrain from childbearing, I would also submit that my refusal of motherhood is alarming—culturally alarming. For as fiercely as I might have conceived of myself as a maverick from childhood, statistics substantiate that my ambivalence (in this instance, a nice word for "hostility") in relation to children is not a private idiosyncrasy, but an altogether commonplace sentiment for my own generation and women born thereafter. Take our maternal emancipation to its logical limit, and all we independent, fulfilled, professionally ambitious Western ladies have a wonderful time, and then, poof! Our civilization disappears. I have enjoyed my life, but I cannot avoid the fact that, as a feminine ideal, there is something profoundly wrong with it.

This is not a small, girly matter. Plummeting fertility rates in the United States and Europe since about 1970 are drastically reconfiguring the ethnic composition, economies, and political balances of power in the West. There may be some poetic justice—nay, vengeance—in the irony that, while motherhood has long been derogated as scut-work by male movers and shakers, Western women's mass exodus from this lowly labor force is hitting elites where it hurts: in the wallet and in the legislature. Impending crises in pension financing, social security, and health care will all be precipitated by women like me. We didn't have kids, and young people pay taxes, bulk up pension plans, and fork over far more in health insurance than they collect.

More crucially still, the disparity between the meager fertility rates in the developed world and the sky-high rates in regions like the Middle East is producing a gathering deluge of immigration—legal and illegal, like it or not. Especially in democracies, this translates into a massive transfer of political power. (Note the push in San Francisco to allow illegal immigrants to vote in school board elections—the thin end of a wedge. And the children of immigrants require no such dispensation in the United States, for if they're born three inches over the border, they vote, period.) The leading reason proffered for inviting more immigrants to the United States and Europe is that "they take jobs that natives don't want." Well, the job that plenty of the natives don't want is, increasingly, having kids.

Few people, of course, resolve the question of parenthood on the basis of what's best for their national economy. Like most people's, my reasons for giving children a miss are personal.

First off, in my own family, motherhood was no route to glory. An aggressive, politically active academic and theologian, my father was forever off at meetings and conferences whose very incomprehensibility left my two brothers and me in awe. Though she did enter the workforce when I was fifteen, during my childhood my mother contributed to the family coffers only by buying day-old Hostess fruit pies and quick-sale vegetables, and her main job was us. Kids have a keen self-interest in discerning who's got the *pow-ah,* and Mother clearly got the raw end of the deal. Dad reaped all the status. Even now, my father is unapologetic about having gone "out into the world" instead of having diligently attended every tuneless performance of my junior-high school orchestra. I can't be critical, either. He did with his life what he wanted to do, and that's what I've done, too. I looked at that model and thought, I don't want to be stay-at-home mommy, I want to be jet-setting dad. And Dad didn't need to be "Dad." He was far more prestigious as Donald.

The other childhood memory that has discouraged me from having children is just my recollection of the day-to-day—what it was like to be a part of a family. Mine wasn't any special nightmare. But like most kids, by the time I was eleven or twelve I was desperate to escape my parents' company. Doubtless, they were often desperate to escape mine. Sure, there were a few golden moments in there, when the Christmas tree lights twinkled and "Silent Night" tinkled in the background and for once no one was fighting. Yet most of the time, someone was getting on someone else's nerves. I don't mean that I feel sorry for myself; rather, that in retrospect I feel a little sorry for my parents. Appeals from the backseat of *Mooo-therrr . . . Gregory won't stop whistling "Yellow Submarine" and it's driving me crazy!*, screaming matches over filling the seven p.m. TV slot with *The Green Hornet* versus *I Dream of Jeannie*, brawls over who gets the corner rooftop piece of Lego—all entirely normal. But that's the point. This is what you bargain for. It always amazes me, the amnesia that seems to descend on so many people when they entertain the cheery prospect of parenthood. Don't they remember what family life was *really* like? And what kind of befuddled optimism possesses folks who, even with proper horror-show childhoods behind them, buy into the fanciful notion that this time around it's gonna be different?

Still, I did reconsider my opposition to motherhood in my early forties. Fertile and healthy, I was then in a stable, long-term relationship, and at last we were in good financial shape. Typically, I reexamined my feelings by writing a novel, a strategy born, if nothing else, of fiscal prudence: at the end of all this introspection I'd fist an advance against royalties rather than a hefty psychiatrist's bill.

We Need to Talk About Kevin is about a woman who, like me, has long been "ambivalent about" (recall, this is a euphemism for "repelled by") motherhood, but who, both to please her husband and to dare herself to do something that daunts her, conceives a son. Either because the boy is congenitally disturbed or because the whole enterprise was contaminated by her initial "ambivalence"—we never know, since our author certainly didn't—the relationship between mother and son is antagonistic from day one. They foster a mutual dislike. As the novel freely reveals in the first chapter, just before his sixteenth birthday, my fictive progeny kills seven classmates, his favorite teacher, and a cafeteria worker at his high school—yes, a Columbine clone, that was the concept. The massacre is and isn't the mother's fault. In any event, this undertaking involved vivifying for myself what it would be like for motherhood to go fatally, catastrophically wrong.

Well, big surprise, I scared myself witless, helping to explain why there are no sticky-faced toddlers clamoring at my feet as I write this. To the degree that I embarked on the novel as a device for resolving whether I would become a mother in the world outside my word processor, I clearly stacked the deck at the outset; the very nature of this dark imaginative exercise ensured a negative result. Nonetheless, writing *Kevin* provided me generous opportunity to contemplate the mind-boggling array of all that can go amiss when one is raising children, and—this much was truly positive—gave me a far greater appreciation for the courageous decision to take the risk, made by grown-ups who read newspapers.

The list of possible disasters is staggering: mental retardation, autism, crippling diseases or handicaps, drug addiction, teenage pregnancy, eating disorders, criminality, suicide—not to mention the standard ingratitude and at least intermittent filial antipathy that come with the territory. Life outside of fiction has bulwarked this admiration, for my younger brother has four children, three of whom have serious medical or psychological problems. The amount of energy, money, time, patience, and tenderness these problems require is inconceivable to the likes of Lionel Shriver, and while my nieces and nephews are lucky to have the responsible, affectionate, long-suffering parents they do, they are also lucky not to have me.

All my reasons for resisting motherhood are selfish. In the main, "selfish" is pejorative. Socially, we need to cooperate, to give as well as take. On the other hand, being successfully selfish at least ensures that I will not burden my fellows. I pursue my own interests and meet my own needs. I do what I want. I blame no one else if the course I have elected to follow fails to content me.

I begrudge my phantom children the time they would require (and deserve); I would rather write more books. I love spending time with my husband, and when we go somewhere, I emphatically do not want anyone else along. Consider this outrageous, but I am making no effort here to be attractive: Were a toddler to impede my ability to light out to play tennis on a summer afternoon when the weather is glorious, I would grow disgruntled and resentful. My income as a writer varies wildly and is perpetually insecure; it's hard enough for me to spring for a new bicycle, and my desire to instead save all my pennies to send someone else to an overpriced private college is immeasurably wee. Small children bore me speechless. I do not want the noise, the mess, the bother, the distraction, the how-the-fuck-did-I-get-here sensation that would inevitably plague me when perched on a park bench and keeping a sharp eye on the swing set.

I would have no expectation that my ungenerous, perhaps even mean-spirited proclivities are of any intrinsic interest to strangers, save for this: I am not alone. What makes my refusal to bear children significant is that my I-hate-garish-plastic-toys thing is becoming widespread. Find my putting a tennis game above the miracle of new life as maladjusted as you like, but I am not a freak. Put together, my small story and many other women's small stories make a giant story.

Cumulatively, petty, self-serving folks like me are transforming the face of Western civilization over the span of a few miserable decades. Though it takes a total fertility rate of 2.1 to replace a generation, the TFR in Germany, where my own ethnic roots lie, is down to 1.3, as it is in Russia. In Italy and Spain—Catholic countries!—the TFR has been languishing at around 1.2 for many years. The TFR of the wealthy developed world is now only 1.5, in contrast to more than twice that in poor countries. Though fertility in the United States is higher at 2.0, that figure is buoyed substantially by the larger families of recent Third World immigrants. Many Americans of European extraction want only one child, or do not want children period. My favorite statistic? An astonishing 39 *percent* of educated women in Germany are not reproducing at all.

Demographers often explain this unprecedented fertility decline in developed countries as economic, and the fact that kids now cost you money instead of tilling your fields is clearly a factor here. But pronatalist government policies with huge tax breaks and the like for offspring are famously ineffective. You cannot pay people to have children. Remember, I said my partner and I could have afforded a child if we had wanted one. We just didn't want one.

Here's where I'm condemned to sound a little highfalutin. I believe that there has been a teleological revolution in Western culture since the 1960s—and not just for women. (I spare you a trip to the dictionary: teleology is "the science of ends.") What has changed since—not coincidentally, in my view—the introduction of the Pill comes down to as profound a matter as why we think we're put on the planet, and what we see as redeeming our existences here.

I think it's safe to say that these days we tend to marginalize the qualities of sacrifice and duty. In the contemporary gestalt, sacrifice is for suckers. We expect our own lives to redeem themselves, in and of themselves, between the goalposts of our own births and deaths. We are less likely to exalt the perpetuation of lineage. We measure the success of our lives in terms of how fulfilling they have been for us. We are less inclined to evaluate ourselves in accordance with whether we have served humanity, our nation, or even our local community. We aim to lead not so much "a good life" as "the good life"—meaning, not good-as-in-virtuous, but good-as-in-fun. As baby boomers like me retire, a great many of us will look back and weigh our choices along these lines: *Did I ever get to France? Have I tried bungee jumping? Did I write that memoir I always planned? Was I fat?* Bottom line: people like me will largely conclude that their lives were or were not well spent on the basis of whether or not they had a good time.

Accordingly, we do not consider children an obligation, because this mentality spurns the whole concept of obligation. (Similarly, staying married is no longer an obligation, either; we stay married only if we want to stay married. We stay married, if you will, for fun.) Children are electives, like that trip to France. Parents who do opt to have families on the basis of the assumption that adoring moppets will improve the quality of their own lives with all that extra love and entertainment are often in for a rude shock. Sacrifice rears its ugly head anyway, and mothers like the one in my last novel discover that children do not necessarily heighten the chances that you'll have a good time.

There's a crude logic to the idea that your own life has to be worth living on its merits, without having to redeem itself by producing another; someone's life has to be worth something for its own sake, or all of human existence is pointless. Nevertheless, a wholesale cultural conversion to the kind of selfishness I'm identifying—an across-the-board embrace of our own lives, and only our own lives, as an end in themselves, to the point that major decisions like childbearing are made primarily in accordance with what is good for us—is disastrous.

Mercifully for everyone, I hope, I'm going to skip the defensive folderol about how I'm not a bigot and how I come from a nation of immigrants and how I love the marvelous, colorful multicultural quilt of this great country. I fear I've already given you reason to suspect that I harbor a pernicious moral barbarism, but at least, for argument's sake, let's take my fundamental decency for granted. We will also take for granted that this whole area of discussion is uncomfortable, fraught with the danger of misinterpretation, fraught equally with the danger of being understood all too well, and thereby of sounding simply appalling.

Yet there are whole university departments now dedicated to the promotion and nurturing of Inuit culture, Native American culture, Asian culture, you name it, and none of these programs is regarded as implicitly denigrating any other people's customs and achievements. So I don't see why I can't say this straight out: I like Western culture. I like Bach and Thomas Hardy and the Chrysler Building; I like Jackie Gleason's performance in *The Hustler*, and even the fourth and fifth reruns of *Curb Your Enthusiasm*. I'm proud of Western culture, and I don't want it to go away. True, we Westerners aren't about to vanish. But our numbers are starting to contract, and in the place of the children we aren't having will be the children of someone else. There's nothing wrong with these visitors-here-to-stay—from Morocco, Saudi Arabia, China—but they speak different tongues, eat different foods, heed different conventions, make and watch different movies, and often hold very different values dear.

You will hear it whispered in some white prosperous circles that "all the wrong people are having children." That sounds racist, and many who express that sentiment are no doubt bigots. Yet one apprehension that helps give rise to this sotto voce exasperation is not necessarily rooted in outright prejudice. While Third World denizens and recent Third World immigrants to the West are still having larger families, these are the very people who can least afford them. Unlike an impoverished mother of six in Bangladesh, I could ensure that any child of mine got a good education, adequate nourishment, quality health care, wide-ranging cultural enrichment, and abundant individual attention. But I won't do it. And I worry that the future is the worse for that refusal. We are in danger of not reproducing the culture that made possible our not reproducing.

Point is, I am a problem. Collectively, I am an enormous problem. Me and my I-don't-want-to-push-a-stroller, I-want-to-play-tennis persnicketiness. Certainly I believe I have the "right" to forgo reproduction, and I do regard my own life as easier for that choice. I have what I want: no short-person company. But I do not like the results, writ large, of myself. I have good genes—healthy, long-lived genes—and were I to have borne children, I'd probably have begotten kids with a knack for language, an ear for music, a sense of humor, and a probing intellect, because that's what my own parents produced. To be sure, both my brothers have had children, and Western civilization will hardly shrivel for lack of Shrivers.

Yet in the case of my husband's heritage, the story is more poignant. He not only hails from an august Ohio family that built an industrial empire producing diesel engines at the turn of the century, but from sequential generations of only children. He and his half-sister are all that's left of the line, and neither has reproduced. The buck—and the baby—stops here. Thus, when I concern myself with humanity's future (with more than one misanthropic bone in my body, I don't always), I have to conclude that my failure to pass these genes along is a shame. On a large scale, multiplied many times, that small shame becomes a tragedy.

Personally, I am grateful that effective contraception has granted me the freedom to lead the life I've chosen, and to not to rise at four a.m. to breast-feed if I don't care to. Socially, it's a fine thing that women like me are no longer obliged to feel that they aren't truly female or worthwhile as people unless they bear children. I do feel like a real woman; I generally feel worthwhile. At the same time, I also regard myself as a prime example of what is wrong with my own culture, and as one more tiny factor contributing to its senescence and decline. If I did have a daughter today, I would urge her to have children. Call it hypocrisy. After all, hypocrisy is all about believing one thing and doing another. I'm glad that I did what I did. But I don't *believe* in what I did. So, as readily as I can defend childlessness for my own sake, I cannot in any good conscience promote it.

Finally, let's close on that standard Rolling Stones dictum that getting what you want should not be conflated with getting what you need. For there are surely satisfactions parenthood affords that, however illusory to the uninitiated, might have made me a kinder, more generous person, even a happier person. Granted, a recent *New York Times Magazine* article cited research documenting that while marriage makes people on average happier, parenthood makes them less so. And you'd think someone like me would seize on that misery index with a smug *Aha!*, as a tool to fortify my self-satisfaction at not having saddled myself with all those happiness-depriving kids. To the contrary: the statistic made me question the whole concept of self-reported "happiness" in such studies, whose definition for their subjects may be too narrow.

Everything that has meant something to me in a profound sense has come to me at great cost. I've taken more than one bicycle trip of several thousand miles, and en route I could hardly have called myself "happy" when getting rained on and buffeted by debilitating headwinds. Yet once a bicycle tour is over, one experiences a tremendous sense of mission-accomplished, and an abiding joy in having undertaken and completed the journey. The same goes for writing books. Writing can be dreary as an activity, even when the words are flowing smoothly, and homicidally frustrating when they're not. And it can be heartbreaking when after all that work you get back curt two-liners about how "I'm afraid I did not find these characters very likable." Especially through the composition of a fragile first draft, I could not honestly classify myself as "happy" in any consistent sense. Nevertheless, my little library of seven novels and counting, in its totality, makes me happy.

True accomplishment—and therefore, from my compulsively Protestant perspective, true happiness—usually entails suffering. Parenthood may be not only an example, but the premier one. Better than the example of novel-writing by a yard. While hackneyed comparisons are often made between authorship and childbirth, raising a whole new person from scratch is a great deal harder than minding an inanimate manuscript, and a great deal more important. So, though it's true that I had books instead of children, that is not a boast. It is a sheepish admission, like confessing that instead of going on a round-the-world sailing trip I opted for taking a bath.

Protective of that narrow sense of my own "happiness," I will never have the opportunity to urge any daughter of mine to have children, or to pass on any other tidbits of unwelcome advice. I will never recognize a familiar obstreperous streak in my eight-year-old, who announces precociously that she doesn't ever want to be a mother, or that she's decided she wants to be a writer when she grows up. I will never have my own adolescent finally discover the world of her mother's peculiar fiction, and treasure a tattered volume of *We Need to Talk About Kevin* on her bedroom shelf. By and large, because it is so alien to me, this ghostly parallel universe does not torment me.

But I think it should.

LUISITA LÓPEZ TORREGROSA

The Life I Was Meant to Have

BY THE AGE OF ONE, before I understood the meaning of motherhood, I knew its physical dimensions, the round, hard belly protruding below my mother's milk-full breasts, the soft lap that had been mine no longer a cushion where I could curl up. When she sat or lay down, I could not so easily crawl on her stomach and sleep there. Something was coming between us.

I was a year old when my sister Angeles was born, a slip of a girl who cried late into the night and held my hand in a tight fist. Angeles and I slept in the same room, she in the crib that had been mine, and I in a small bed beside her. Two years later, my mother was pregnant again, a condition I came to identify with the wondrous belly expanding beneath her loose cotton dress and the pastel ribbons she wore in her hair. For reasons that only a child can invent, I believed that the ribbons made her pregnant, that the big belly and the ribbons came at the same time. They were symbols to me, an announcement.

My parents had six children together (my mother had a seventh with her second husband). Despite her large brood, my mother, a lawyer, happily worked all her life and didn't find being a housewife especially rewarding. When she stayed at home for a prolonged period, during her one month's vacation, for instance, she would bite her nails to the quick, lie in her bed reading books and staring off into some distance—or go around the house dusting the furniture madly, showing the maids the right way to do it. If she were alive, she would not deny this, that she didn't like being a stay-at-home mother. We, her children, knew that as often as she said that we were her life, she had other lives that filled parts of herself that were left sadly empty when she didn't have an office to go to, a newspaper to read, a poem to recite, a dance card filled with admirers.

My father, a doctor, a traditional man with Old World ideas (affairs, heavy drinking, running the house as his fiefdom), had little time for us. Together with my mother, he gave us everything—good schools, closets full of clothes, a proud name and position in the upper-class society of Puerto Rico, where I grew up. But he did not put on his slippers when he came home in the evening, gather us at his feet, and read us fairy tales. He missed all of the milestones—graduations, birthday parties, piano recitals. He was a shadow, a very large shadow, and his voice and rules were our command.

On the day I began menstruating, at age thirteen, my father came to my bedroom and in a chilling, somber tone told me that a woman has nothing bigger to give the world than children. My father would drive that idea home more forcefully years later, when I was sixteen and wanted to go away to college. "You are a woman," he said. "You need to know how to cook. You need to get married and have children." My mother, by now divorced from him and furious that he had betrayed his earlier promise to pay for my education, found an affordable U.S. prep school and then an all-girls' college in South Carolina, and sent me there. But even she expected me to return home and marry. As much as she showed independence by practicing law when few women of her generation could or dared, she believed that being a wife and mother were the most important roles in a woman's life. "A woman is not complete without a man, without children," she said often, and she truly meant it.

How then, with all this pressure and my parents' expectations, do I find myself in middle age without children? Was that born in me? As a child, my sister Angeles, who grew up to be the most beautiful and seductive of all of us, would spend days with her dolls, dressing them, cutting out paper patterns, sewing them clothes, cradling them as if they were flesh and blood, while I was out with the boys, playing baseball, or dressing up as a cowboy, pistol holster at my waist, Roy Rogers hat on my head. Had I already decided—or had the decision been made for me—even back then?

I found boys, and later men, sexy but slightly scary, with rough hands and coarse lips. I was in prep school, maybe fifteen, when during a holiday in Long Island, I was set up on a blind date. We went to a drive-in and sat in the backseat of the car. My date (buzz cut, heavy arms, football shoulders) pulled me to him, jammed his lips against mine, his tongue furiously probing my mouth. I nearly gagged. He was breathing hard and his hands grabbed my breasts. When I pushed back, he laughed. When I started to cry, he slammed the door of the car and left. I was mortified, so distraught that I did not leave my bed the next day. It was there, for the first time, that I had a terrifying thought: What if I didn't like men at all?

Away at college in South Carolina, apart from my family and the pressure to follow the customs and traditions that were bred in me, I tried to enjoy dating. I listened both shocked and entranced as girlfriends recounted tales of boys and fondling in dark rooms. I was curious—what was that like? Sex—the *idea* of it—was repellent, and yet strangely exciting, and I couldn't understand the incongruity of those impulses. I couldn't understand what was keeping me from committing myself to a boy, from giving myself totally to a man. I was young, though, and I thought: there will be time for all of this. I will eventually learn to like sex. I will eventually fall in love.

And I did, in a way, when I was eighteen. But it wasn't with a boy. It was with a girl who lived across the hall in my college dorm, a girl with soft jet-black hair and dreamy green eyes. One day she asked me to rub her back. Touching her skin, feeling the goose bumps on her neck, kneading my fingers into her muscles and letting my fingertips come close to her bare breasts, hearing her sighs and noticing her eyelashes fluttering, her eyelids closed, I was mesmerized, lost. I had had crushes on girls, which had left me sad, confused, and ashamed.

But I had not felt anything quite like what I felt that afternoon, innocently touching her flesh.

One late evening, she came into my room. She didn't say a word, and she made love to me. We became lovers, insatiable, crazed, and reckless, partly thrilled by our secrecy and afraid, too, that any night the door would swing open and we would be discovered. Our affair lasted two years, even after we had finished college, but we were destroyed eventually when her parents found out. She left me, and months later I heard she had married. Hoping my feelings would change, too, I continued going out with men. But my experience with them repeated itself in many variations over the years: flirtation, desire, repulsion.

My mother was in her forties when she gave birth to her seventh child. She was remarried, living in Bryan, Texas, and this child was a gift to her new husband—or, perhaps, an obligation. Either way, she believed that giving birth wasn't something to be avoided (she never used birth control). She had done it so often, so easily, she said, that it was not so different from breathing.

I didn't believe her. I was twenty years old, and my mother's only companion on the day my half-brother was born. When her water broke, her husband was away on a business trip. Mother called her doctor and picked up the small valise she had ready for just this moment. I drove her to the hospital, and she checked in calmly, as if she were booking herself into a hotel. She walked steadily, her round stomach hardly protruding from her black-and-white maternity dress. I marveled at how collected she was.

My mother's doctor allowed me into the labor room, and I stood by her bed and held her hand as tightly as I could as she moaned and twisted with each violent contraction. Her face was contorted, her lips dry, her eyes moist and wild. Her skin looked and felt like candle wax. The sheets were soaked with sweat, her hair a wet mop, strands sticking limply on her face and neck. Her breath was foul. She looked as old and haggard as I had ever seen her, but I forced myself to stay calm, the quiet that comes to me when I am most afraid. I wasn't scared that she would die. I was afraid because she didn't look like my mother. As I watched her writhe in pain and as I cringed at the screams coming from other women in the ward, I swore I would never subject myself to this particular hell.

"You forget the pain," my mother told me later, as she lay in her hospital room, looking exhausted. She was forty-four years old and had a new boy. When they brought him to her, she tried to smile, but she was half-drugged, worn out, her uterus bleeding and torn. Soon her husband arrived, and flowers filled the room. The baby was swaddled in a tiny blanket, and we all remarked that he looked just like his father. Hours later, Mother was out of the bed and walking about, and the cries I had heard coming from her, and the face of pain I had seen, had vanished. She was freshly dressed and made up and looked young again. She had a new son. That was the only thing that mattered.

The only time in my life I wanted a child was when I was twenty-four years old, living in North Carolina, and having an affair with a married man. He was gentle, bookish, and not particularly good-looking. I was rebounding from a couple of dead-end affairs (one with a woman, one with a man), and I became attached to this funny storyteller who treated me like a princess, who was in love with the twists of my mind and the shape of my body. He knew about my love for women and it didn't matter to him. I felt safe.

Over the months that we saw each other, I began to believe I could have a child, if it was with him. I romanticized motherhood and savored the image of me carrying his baby. This motherhood spasm went on for weeks. I confided in friends—would they stand by me if I had a baby out of wedlock? They were thrilled.

Then, one evening, I snapped out of it. We were at a party, exchanging those furtive glances that secret lovers give each other, a way of claiming ownership. We had been drinking, but no more than usual, and left the party in separate cars, knowing we would meet later at my house. There, after a swig of scotch and a kiss, he looked at me very seriously and told me he wanted to divorce his wife and marry me. I was stunned. He had a wonderful wife (I had met her once) and two children. It had not crossed my mind he would ever leave them. I stared at him, touching his face, holding his hand. The reality of having children, of having a husband, was too real, too unpleasant. The fantasy crashed just like that. The truth was I did not want to give myself totally—to him, to a man, or to a child. We can't, I said. I can't let you break up your family, and I can't marry you. I didn't tell him I couldn't marry him because I knew that eventually I would fall in love with a woman again. But he understood. He knew that was the reason.

Eventually, my mother gave up on my becoming a bride and mother. It wasn't easy for her. But she came to understand that my life had taken a different direction, and there was nothing she could do about it. We never fought. There were never any awful scenes. We simply grew older, and she and my sisters stopped asking questions. They knew I wasn't going to bring home boyfriends. My mother, who I know thought about it and wondered about it (and punished herself about it, probably) was as loving to me as she was to the daughters who gave her grandchildren. But though we never talked about her disappointment in me, it was always there, a certain tension that at times was palpable. I had turned my back on the trappings of marriage and children that she so believed in. That conversation which we never had was the one we should have had, but we were too delicate, too afraid to broach the subject of sex, a topic my mother would gingerly avoid. She would not have wanted the details.

For most of my life, children were strings attached to marriage, to a man, but now just about anyone can be a mother—married, single, divorced, gay, old. If these options had been available to me when I was younger, would I have chosen to be a mother on my own or with a lover? I don't think so. I was not made to be a mother. I can't explain how I know that, but I do, in my core. I know this now even more strongly than I knew it when I was twenty-five or thirty. With children, I wouldn't have been able to travel the world, wouldn't have lived in Manila, where I experienced the most intense years of my life. I probably wouldn't have been able to put down my roots in New York City. I might not have written books. For me, children likely would have taken the place of writing, or they would have distracted me from the thing to which I feel I must give everything. I would have cheated them. Friends ask, don't you regret not having children? No, I tell them. This was the life I chose, the life I was meant to have.

And yet I understand the need, the desire, to give birth to flesh and blood; I understand that creation. A colleague brings her first-born to the newsroom where I work, and her face glows—she is in love. I look at her and I know that nothing in her life—her job, money, even perhaps her husband—will give her the joy that her son does. I see the same look on the faces of my sisters, all of whom have children, even as they balance motherhood with their own independent lives, sometimes with frustration but never with regret.

Most of my life I believed that writing books would be my way of having children. So many writers believe that, and it's partly true. Books are creations, unique to themselves, all your own. They, too, live after you. But now that I've written a book, now that I am starting the next one, I'm not so sure that writing will ever give me the complete fulfillment I hear in my sister Angeles's voice when, touching her son's hair, she says, "My boy."

In her home, there are two photographs placed side by side, one of my mother holding Angeles when she was less than a year old, and one of Angeles holding her son when he was a few months old. The photographs are nearly identical, similar poses, similar expressions, similar smiles on their faces. I see right away the meaning of motherhood reflected in those two photographs: they are frames of continuity, a kind of immortality, the flow of generations.

LESLEY DORMEN

The Impossible Me

I AM NOT, NEVER WAS, AND NEVER WILL BE A MOTHER. No more someday, what if, you never know. I'm in my mid-fifties. I know. I didn't mean for my life to work out this way. Without ever exactly choosing this outcome, I must have been choosing it all along. Otherwise, how could this have happened? The question asks itself whenever I see certain little girls on the street, stubborn beauties with tangled hair and sticky hands wielding their considerable power over mothers and doormen and shop clerks and dogs. That's exactly what I miss, what is now impossible: little girls between, say, three and eleven.

I miss my heart expanding to make room for the passion I imagine feeling for such a child. More missing, in descending order: my marital team of two being made that much stronger by the addition of one or more. I miss whatever edge might have been worn off the contemplation of my eventual non-existence by knowing a small piece of myself would continue on in my absence. I miss the narcissistic kick of seeing myself reproduced in another human being. I miss sharing the camaraderie of a certain breed of mother, the wised-up, no-nonsense kind I think I would've been. I miss needing a copy of *Goodnight,*

Moon. I miss the shock of having all my preconceptions and fanciful notions about children overturned by real motherhood, the state itself.

As it is, I experience the entire range of maternal feeling in miniature several times a day in my child-choked Greenwich Village neighborhood. Coming out of a florist's shop recently, I stopped to watch a little girl around three marching along the sidewalk. I felt delight in the confident curiosity of her stride, then concern as seconds passed and no adult materialized. I was relieved when I heard a woman down the street who I supposed was her mother call out to the child to hurry along, then filled with suspense as the girl hesitated before continuing her march in the opposite direction. When she turned back, and I saw what she saw—no mother, as the woman was now obscured by other pedestrians—I was scared. "Mommy!" the little girl wailed despairingly. And finally, I felt lonely and excluded when the child ran into her mother's arms.

As the drama played itself out, I watched not only it but myself watching it, identifying with the child as well as her mother. How far up the block could I allow my own little girl to wander before being compelled to scoop her back up? And as I tried to balance those two competing parental values—protecting versus encouraging independence—I was made to recognize all over again that this was a calculation I'd never have to make. Only moments had passed, but sometimes not being a mother feels like a full-time job.

I can hear my mother, who died seven years ago, saying in her warm and lovely voice, "Having children is the one decision you will never ever regret." I was in my early thirties. We were having tea at the Plaza. It was a beautiful spring afternoon, with my mother's childhood friend Eileen visiting from Cleveland. The three of us were catching up.

"Having kids was the best thing I ever did," Eileen echoed over her dainty cup of Earl Grey.

I realized that my mother had meant the tea as an intervention. Despite my "advanced" age, she knew children were the last thing on my mind.

I'd come to New York directly after college in the Midwest—

unmarried and with no particular ambitions. I needed a job, of course. But I didn't want or know how to go about getting a career any more than I wanted or knew how to go about finding a husband or having children. My mother's two divorces, financial insecurity, and various other fractures hadn't given me much practical sense of how to proceed, never mind the necessary confidence or conviction. All I wanted was a best friend, enough books, a boyfriend, pretty clothes, distraction from the prickly fear always nibbling at me.

So I spent my twenties hopscotching from one dead-end editorial job to another, much the way I jumped from one wrong lover to the next. Restless and unmoored, I moved to a small town in Connecticut for a few years, thinking I'd find whatever I was searching for there. Depression found me instead. Although I showed up for work, paid the rent, maintained, however raggedly, important relationships, I spent my thirties—the last decade of my fertility—deciding every day whether I wanted to live or die.

That afternoon at the Plaza, I was touched by my mother's good cop/good cop strategy. I was touched she even had a strategy. I was also indignant. How could these standard-issue Jewish mother warnings be of any use to me? My mother and her friend had had three failed marriages apiece. Each had been a creative businesswoman at heart but found only minor or intermittent success. They were women with finer taste in clothes than in men, women with more style and charm than opportunity or focus. How could they advise me? I wasn't thinking about motherhood because I was trying to build a life worth living for. Tell me how to not be afraid to live. Tell me how to love for three months without grieving for three years. Tell me how to become the writer I've just begun to want to be. Then tell me how to have a child to complete all of the above. Not that they could. Not that I'd have listened if they had.

What I also remember from this time is a despairing summer weekend. I was at the end of a disappointing love affair and had just found out that I was pregnant. I'd scheduled an abortion for the following week. I was unequipped financially and emotionally to consider having a child on my own. I had no doubt about this decision. But I remember noticing on that lonely summer weekend how even the promise of a child seemed to add gravity and value to my experience of myself, how knowing I was pregnant seemed to ameliorate my anguish just the tiniest bit. I felt larger than myself, a little more important. I had the abortion.

Recently, I dreamed of two big-headed baby dolls. In the dream, I'd forgotten to feed the dolls (or were they babies?). My confusion and fright and shame were so intense upon waking that I was reminded of a painful comment a friend made a few years back as I attempted to serve crumbling pieces of birthday cake at a dinner in my home. She laughed and said, "You don't have a maternal bone in your body." What? Was it true? I must have thought so, judging from how wounded I felt. Silently, I mounted a defense: Loving wife! Compassionate teacher! Was there ever a dog I didn't stoop to hug? Empathic and loyal friend. Soup maker! (Yes, it came to soup, that's how bad it was.) No, I decided. Not true. I would've made a good—well, good enough—mother, but what if believing oneself to be a good mother is like having a good sense of humor? It's something, as Nora Ephron once observed, that everybody thinks they have.

Not being a parent, I've lost out on knowing the fuller story of my character. No one will ever count on me in so final a way as a child counts upon a mother or father. No sons or daughters will ask me to cough up guitar lessons, college tuition, mortgage money, or bail. I don't know how well or poorly I would have handled the pressure, the singularly demanding responsibility of shepherding a child into adulthood. Might motherhood have made a (better) woman out of me?

There's another side, of course, to the reflexive questions, to the wallpaper with its endlessly repeating pattern of mother/not a mother that lines the inside of my mind. There's the sense that I've gotten away with something. Frankly, I wonder how anyone survives being a parent. I'm not sure the joy and hilarity, the privilege of living alongside a growing child, could really compensate for all the work and worry and risk. I'm a little ashamed to say it, but there it is. I feel almost criminally free to be absolved of so much toil and angst.

I'm free to read and write mostly—the passions I did choose, in my own haphazard way. As I write this, for example, I'm getting ready to spend seven weeks working on a short story collection at an artists' retreat in New Hampshire. Certainly there are women who figure out how to arrange their lives so as to both make art and care for children. But it's taken me a long time to recognize myself as a writer and even longer to become the kind I want to be. I didn't begin writing with the seriousness of purpose I'd imagined and craved until I was almost fifty. I'm still at the early, getting-to-know-you stage of fiction writing. So, the freedom to keep honing my craft, to follow wherever it leads, is precious to me.

What if back in my twenties I'd been asked to choose art or children—which would it have been? If one of my female writing students asked me today for help with that same choice, here's what I think I'd advise: Children. Even though I know that life can be pleasurable and profoundly rewarding without them. Even though I happen to pity the artless. Even though I think the world could do with more art and fewer children. I don't know why this is or even if I'm capable of understanding my own contradictory answer.

I see myself as the end of my thirties came into view, in my gynecologist's office for my annual exam. "How many sexual partners do you have—one or more than one?" my doctor asks me after the checkup. This was in the early years of AIDS.

"Less than one," I say. I thought my fertility would last forever. Fertility wasn't part of the national conversation yet. Perhaps if it had been, I'd have focused on my future with more seriousness. In fact, I remember telling that doctor at a later appointment—I was forty by then—that I felt quite fertile.

Also, I'd become accustomed to thinking of myself as a late bloomer. Why should motherhood be any different? Depression had obliterated all desire for so long that when it finally lifted and the world reverted from black-and-white to color, I believed that everything was still possible for me, including motherhood.

"Don't wait too long," my doctor cautioned.

Nobody was talking much, either, about the varied ways of becoming a single mother. Although I had one single friend who'd adopted, and another using a sperm bank, I never considered those options. Love and work, that's what occupied me. I was supporting myself as a freelance writer by then, but I still hadn't found a stable romantic relationship. These were the quaintly ironic years when single women past a certain age, it was said, were more likely to be the victim of a terrorist than to marry. I still wanted marriage very much, more than I wanted a child. Or maybe that, too, was a failure of imagination. I couldn't imagine one without the other.

When I finally walked down the aisle at forty, it was with the understanding that I'd get pregnant right away. But that relationship was ill-fated. Our love affair had been on and off again for many years, and the wedding turned out to be a complicated means to its conclusion. We separated within the year. I hadn't become pregnant.

Now my fertility really did grab my attention. Time was running out, I knew. But I didn't so much feel a specific longing to mother as to have as rich a life as possible. Work remained my paramount concern. I had taken a few fiction-writing workshops in my late twenties, but I'd put away my dreams after completing a handful of short stories. I hadn't understood how much I had to learn and that the complimentary rejection letters I received were part of that process. Anyway, I needed every speck of energy I had just to survive during those years. By the time I was forty, though, when I read work I admired, I thought, I want to do this. Can this be me? I was also watching new mothers, those dazed and love-struck beings, and asking myself the same question: Can this, still, be me?

On one of our first dates, my future husband told me not only did he not want more kids—he had three handsome, almost grown sons from his first marriage—he was unable to have more. "But why didn't you insist?" one good friend has often said to me, ticking off all the available methods for a woman who wants a child to have one. "Didn't he realize how much you wanted a child?"

I tell her, and myself, that having met a man with whom I thought I could live my life I had to take him as he was. My husband didn't owe me a child, I protest, and if I'd been thirty when we met instead of forty-two, I might have thrown the dice again and moved on. Yet maybe I was simply afraid to put the relationship on the line. What would have happened if I had? I just got up from my desk and asked my husband that question. He pondered it for a good ten seconds and told me that he probably would have walked. He'd ended a relationship before ours for that reason. I probably was too afraid, suspecting that dismal outcome. But again, I have to answer to my own ambivalence. I just didn't want a child enough.

When I'm with the children of friends, I'm mostly charmed and curious. I read articles about child development and parenting. I'm thrilled to go to child-oriented events. And though I occasionally feel regret for what might have been—observing the practiced ease with which my husband slings an infant over his shoulder or amuses a toddler—the feeling is fleeting. The absence of a child lives in me, but I don't dwell near those moments or in them. I have to conclude that it's impossible to deeply mourn a child I've never had.

When my mother got sick, I took her for an MRI. I stood alongside her as she disappeared inside that noisy tunnel. I held her foot and thought, who will hold my foot? But I can't say the question of who will care for me when or if I'm old and ill really bothers me. What I'm wistful about is my album of family photographs, my grandfather's eyeglasses, my mother's dessert plates. What will happen to these things when there's no one left to treasure them?

I have an old photograph I bought at a flea market shortly after I first came to New York. It's one of those typically posed tintypes, of a young girl around the turn of the last century. Except this girl isn't unnaturally frozen. Her hair is unruly, and the photographer has captured her just as one hand is flying up to cover her face. Who was this girl? Was she feeling melancholy or frightened, impatient or playful? What was her life? She's been with me for a long time now, but I find the mystery of her as compelling as ever. Whom did she care for and who cared for her?

PART TWO

On the Fence

RICK MOODY

Bloodthirsty Dwarves

I FIRST MINTED THE COINAGE IN MY LATE TWENTIES. I was in one of those difficult relationships where one person wants one thing and the other wants to put it off for a decade. Indeed, this impasse appears more often when you're in your thirties, at least according to the conventional thinking. But I got my start earlier. The woman in question was from an unusual family. Her dad had been a child in hiding during WWII. His own parents, who'd managed to avoid the camps by sheer courage, had been killed not long before the armistice. In Poland. He was one of the bravest, most interesting people I've ever met, but he was not effusive as a dad. He was the opposite of effusive. Her mother, meanwhile, was a frantic Irish Catholic woman who seemed forever to be renting out portions of her property up in the suburbs, and in the process, scheming to avoid detection by the local zoning board. Of course, these parents were long since divorced. In the wake of their union was this driven, ambitious, jealous, headstrong, apoplectic, beautiful woman—my girlfriend. Let's say her name was Irene.

Irene's black rages were legendary in her family. I went home with her for Thanksgiving once. We'd not been in the front door twenty minutes when she told her mother to fuck off, in front of everyone. Things did not improve later. Once, Irene came out of the bathroom of the apartment we shared and accused me of avoiding changing the toilet paper during the entirety of our relationship, going on four years. There was a lot of shouting. It was an erroneous and unsubstantiated charge, but that wasn't the point. The point was that Irene could go completely out of control faster than a Lexus can get to sixty mph, *and it was obvious that I didn't give a shit about our relationship or her, and so she was going to walk out the door again and I shouldn't expect to ever see her again!!!*

In the middle of her oscillations, in the days between up and down, she was pressing pretty hard on the issues of marriage and children. Normally, I ducked these subjects, subjects that made me anxious, and I'd done pretty well ducking them for many years. Well, maybe it's more honest to say that I had not been a very dependable person in my middle twenties, with substance abuse problems, mental health problems, et cetera. Marriage and children were not in the front window of my consciousness. Additionally, there was my own parents' acrimonious divorce. And their unconventional child-rearing practices in the early seventies. I mean, I was no angel. No one with insight into character would have wanted me to marry them when I was young. It was *easy* to avoid the subject of marriage and children back then. But by the time Irene and I had been together for several years, conversations about marriage and children were like the category-four hurricanes in the Caribbean of our relationship. Two or three times a month, and there was no predicting when, Irene would get to looking like I had just stolen money from her or read her diary, and she would start in on me about why I didn't want to *commit.*

The obvious answer to all this would be why *commit* to someone who could easily inflict puncture wounds upon your person for not changing the toilet paper often enough? But that would not be the true answer. The true answer was that, at least for me, I could never (and can't still) enter into any arrangement or partnership or joint activity or anything with another person when force is exerted upon me. The more force is exerted upon me, the more recalcitrant I get, and I don't care if my recalcitrance is self-destructive. On the contrary, it's even better if it's self-destructive, in a way, because then the person exerting the force can see the high cost of their pressure. Look, I'm coming apart! And you did it! The big enough lever that might move the entirety of the world could not move me if I felt like there was coercion involved, and I was happy to bleed or waste away for the idea, if this was what was required.

One night, Irene was giving me an earful about how we really needed to get moving on these *long-range planning issues* and didn't I think we would be good parents, et cetera. To which I replied, "You know what? I think children are nothing but *bloodthirsty dwarves.*" It was rhetorical savagery. It put a real damper on the conversation, it put a damper on our relationship, it put a damper on couples therapy, where we'd been going for six months or so, it put a damper on our home life.

I had this theory, see, which I articulated in reply to the dominant ideology of the *commitmentphobic male*, and my reply was the ideology of the *contingencyphobic female.* According to my theory, in the cauldron of feeling like they were supposed to want marriage and children because that was what the social engineering of American culture wanted from them, especially in the Reagan-Bush years (of which we are speaking now), women projected their own ambivalences about commitment and children onto guys so that they wouldn't have to feel it themselves. And what was it they were not feeling? They were not feeling *contingency*, the nature of change in all things, the acting-out of entropy among lovers and in homes and families. *Contingency* meant that the only commitment was in the present moment, and, I argued, I was always committed now, I loved Irene now, I loved love now, and what was this future but a sequence of present instances? Irene, and the armada of women's magazines spokespersons who acted as a bulwark in the creation of the archetype of the *commitmentphobic male*, were incapable of feeling just how frail and gossamer was this romantic bond between people. All things metamorphose into their obverse. All things change and merge and end. So, commitment is shallow, is a kind of binding lie, nothing more.

Reasoning like this drives people off. It's the same with the "bloodthirsty dwarves" line. And that may have been what I was doing anyway, driving Irene off. After a week in Maine, one August, when (she later told me) she cried every night while I was asleep, we agreed to part. I can't say that I learned much from this experience, either, because when I finally met someone new, six months later, I was just as resistant to the traditional duties of marriage and children as I had been with Irene. And things would probably have gone on in much the same way, with me arrested at a certain point in my thinking about domesticity, and coming up with really baroque rhetorical flights for anyone who tried to persuade me otherwise, if not for one of those really wretched and horrible tragedies that always seem to temper your thinking and take you places you never thought you'd ever go. For me, the tragedy, as I've elsewhere written, was my sister's death, and the specific lesson was not so much in that dark day and its unthinkable losses than it was in what remained afterward, namely, my niece and nephew.

Since my niece was five and my nephew seven in 1995, when my sister died, it seems like I must have had some kind of relationship with them before then. There is videotape and photographs to prove as much. They were cute kids, devious and diabolical in ways that I believed all children were, with intractable characters and unreasonable wishes (had to be pasta every night, don't even think about putting a vegetable in front of them, et cetera), but they were related to me, and that was faintly interesting, I supposed. They turned up with my sister during the summer, when we were all together on the coast, and they were sometimes funny, and sometimes they got in the way of proper conversation. That was about what I thought of them.

The fact was, however—and there's really no delicate way to put this, so I'm just going to say it—that my niece and nephew were *with* my sister when she died. Since she died quite precipitously (I'm leaving out the details now, so as to spare the principals), without forewarning, you would be right to infer that my niece and nephew observed something that they should not have had to observe. If the fates were not governed by *contingency* but were more orderly and just, my niece and nephew should have been eating chicken fingers or watching television or putting on their pajamas.

So, the kids had seen something they should never have seen, and the question that follows is *what do you do about this?* What do you do, as a thoughtful person, for kids who have seen too much or experienced things they should never have experienced? The first thing you do is you walk around with them and play games, if that's what they want to do, and you go see movies that you would never otherwise see, including films starring Adam Sandler, and you ask after what's going on at school, and you make clear all the problems you had growing up—and if you're me, that's a fair number of problems—just so that your niece and nephew will not feel like they are in this alone, that they are the only people who have suffered this much, and then you listen some more, and then you listen long after you might otherwise stop listening, and then you listen hours beyond that, you listen because listening is gentle and also because it's dependable, and then you listen just because it's the best way to express love. It may be that one day my niece and nephew are the ones who need to tell this story, instead of me, and I'd like to think that when that day comes I will be happy to shut up at last so that I might listen anew. Hopefully their version will make clear how years can pass this way, very good years.

Which is to say that I felt in this tragic part of the story a fairly dramatic change in the way I related to the "bloodthirsty dwarves." I think I had already occasionally been the kind of adult of whom people say *You'd make a good father*, mainly because I had and still have access to the part of myself that is interested in *play*, because this is a good wellspring from which to draw in the matter of being creative. I am not afraid to get paint all over myself, nor to dig up worms and then rebury them. I don't mind chasing rabbits or repeating an activity seven thousand times. I'll do what it takes to spend time with a kid. After my sister died, what happened, emotionally speaking, was that I started wanting to do more with the kids around me. I wanted to be around them more often, I wanted to be a resource for them, I wanted to be in their lives. Didn't matter, particularly, whether they repaid me for this, didn't matter if they cared, if they respected me or not.

In a way, this need was selfish. Because it was a route through the loss I felt about my sister. And it was a way to slay some time when I didn't really know what to do with myself anyway. Hang out with the kids, keep your mouth shut. Repeat. In a bit of awful timing, my sister died two weeks after my younger brother's wedding. Not coincidentally, therefore, my brother and his wife began trying to have a child not long after her death. They were fortunate and had no trouble conceiving. Thus, the second batch of nieces and nephews (there are two, as of this writing, and a third on the way) was soon hatched upon the scene, and that only increased the kind of satisfaction that I seem to get, whether I want it or not, from a generous participation in the growing up of the next generation of my extended family.

That I happened to be involved with another woman during this period of delight in children, a woman alluded to above, probably helped. That the woman in question is very easygoing and relaxed and joyful around children has not hurt. I think it's fair to say that she loves my nieces and nephews as much as I do, and has been as involved as I have been in trying to provide for them sturdy, reliable, and relatively stable adult perspectives. It seems now that getting from this spot, the spot of being a good uncle, to the next spot, where you are getting married and starting a family yourself is, self-evidently, just the proverbial hop, skip, and jump.

I am, therefore, a married guy, now, though I long suspected I would never be. And I am, in fact, on my honeymoon in New Zealand, with the sun coming up over the Pacific Ocean out the hotel window, and my wife asleep in the next room. And we are already agreed that we will have children, and they will come or not come according to the *contingency* of God or the fates or whoever is in charge, which *contingency* remains the one reliable thing in a world of considerable instability.

Does that mean that I'm a convert? That I think children are just so *cute,* and I love those *little shoes that they wear*? Does it mean that diaper-changing is just one of those moving, intimate parts of life and it really doesn't smell *all that bad*? No, I'm not a convert; yes, I'm still a skeptic; yes, the world is overpopulated and I don't particularly want to contribute to that grim trend (which is why I have committed, in theory, to only one child: net loss to world population—one person); yes, we live in dark political times and it's scary to think of bringing a child into the world in the midst of it; yes, I have work to do, novels to write, people to see, and I don't want to sit around talking baby talk, yes! I hate the word "poopy"! The word "poopy" will never cross my lips! And I am not going to sit around with rich shopaholics on the playground and talk about stroller design! I will never own a sport-utility vehicle! I understand these arguments, these disinclinations, I understand my friends who say that they will not *reproduce*, and I think their decision is fine, and appropriate, and some days I even think back on that time in my life with a warm nostalgia, the time when I had few complications and knew a lot less about death.

But what I like now is responsibility. What I liked when I was young was liberty, and I was willing to draw rigid lines in the protection of that somewhat narcissistic liberty, and what I like now is being able to show up, to the best of my ability, and being able to create some lasting sense of what I did with myself here on earth, even if all I did for the humans around me is listen. Even if that's the only really lasting part of it, that's still good, and that's a kind of creativity. Every mother who had her kid at nineteen knows that what she has done is creative. Each day of her life, she gets to drink deep of the satisfaction of knowing that she has made a great creative sacrifice. I don't see why I can't be a part of that, too. So here goes.

REBECCA TRAISTER

They Will Find You

I'M TWELVE YEARS OLD, lying in my bed, hoping to get to sleep before my father turns his classical music off, my mother stops correcting papers at the dining-room table, and my home falls dark and silent. I suffer from night fears and hate being the last person awake in the house, responsible for battling any monsters that might invade it. I'm soothing myself to sleep by fantasizing about next year in the seventh grade. I've just taken the tour of the local junior high, and its corridors and lockers have spoken sharply to me of grown-up life. Once there, I know, life will be very different. I'll have boys kissing me, feeling my breasts, making out with me under bleachers. It feels thrilling and embarrassing, and as curious as I am, part of me wishes I could resist it. But I know that this is inevitable, unstoppable, this change from little girl into woman. It's out of control and unfathomable—like my body, which is already an embarrassment of growing breasts and widening hips. And I realize that as awkward as all this is, it's simply how it happens. Going to junior high is the first step in my transformation from ungainly duckling to perky teen, college student, sophisticated adult, married woman, and mother. I can't choose to get off this ride, which will likely dump me where my mom is now, working at the dining-room table while a squad of neurotic children fret in their beds.

I would say that I have always known I wanted to have children, except that, of course, for the first couple of decades of my life, the thought didn't cross my mind. I never played mommy to my dolls and was largely unmoved by the pregnant women who crossed my path. I hated babysitting. I wasn't ever one of those girls who gave a lot of thought to what she would name her offspring—even when pressed by friends eager to share that they had committed to "Holden." I devoted much more of my imaginative time to what kind of speech I might give at the Oscars, and who might take me and my statuette to bed afterward.

But if romance were a kick-ball team, some deep part of me feared that I was going to be the last one sitting on the ground. I couldn't think what actions I could take to make it any different. So I took solace in the notion that if I just kept still, life would pick me up and take me to the adult world, where love was inescapable, sex was easy to master, and babies were born without ever rupturing the space-time-career continuum. While I agitated about whether my grown-up self would be dramatic and successful and skinny and well dressed—qualities over which I imagined I could exert some control—I basked in the knowledge that love must be an area that simply fell into place on its own, kids hot on its heels. Because I was flummoxed by it, it surely involved an alchemy that would transform me in time, beginning in the seventh grade.

I'm eighteen, about to head to college in the Midwest. I hate my parents, but am still comforted at night by the sound of my dad's classical music. I know—from my older friends and from the brochures—that life is about to change for me. I'm anxiously anticipating the alterations, since high school wasn't exactly the Breakfast Club I had imagined it would be. I never got those corridors of lockers or amorous nights under the bleachers at the local high school. At the last minute, I decided to go to a private Quaker school, where we smoked a lot of pot, drank a lot of beer, and learned a lot about the Peloponnesian wars, but didn't do much dating. There were only a handful of actual couples in my graduating class of eighty. People had sex—though I didn't—but it was of the couch-in-the-basement-after-everyone-else-has-passed-out variety rather than the I-love-you-but-do-you-think-we're-really-ready variety. I'm feeling a little sad. Not sad about my virginity exactly. But sad because I wasn't brave enough in high school to be one of those people—the kind that had sex on the couch. I was chicken, and now I have missed out on dating as a casual sport, missed my chance to get to college with the sexual proficiency I'd expected. In college, I know, there will be sex and more beer and more pot. And I know that a lot of it will be casual, but that the stakes are ratcheted way up. College is likely where I will meet my husband. All through high school people have told me, "Wait till you get to college." It's where most adult couples I know first sniffed each other out, where they met and mated for life. Realistically, the next time I live in my parents' house, sleep in my childhood bed, I will not only not be a virgin, but may have already met the man I'm going to marry, the man I'm going to have my children with.

My parents met in college, married right afterward, and then waited almost a decade to have their two children. It's a model I'd like to emulate. I hope to have a lot of years with my husband under my belt—to have a relationship that is about the two of us—before we bring other people into it. I figure it's going to be tougher for me to be a talented wife than it is a talented mother. After high school, I do know something about myself that I didn't before. I know now, in a calm and unworried way, that I will be a good mom. It's partly because I know my body better now. It's made for childbearing. It's tough and sturdy, and the period I got before most of my friends—and it has stayed regular while theirs waned with anorexia—tells me that I am one of those girls who's simply, plainly, calmly fertile. Loath as I am to admit it, I know that I will someday take on the characteristics of my own mother. Placid and eminently capable, my mother doesn't go in for a lot of self-doubt or neuroses. She works full-time as an English professor and gets groceries, cooks fresh for four, washes dishes, and does a load of laundry every single day—not only without complaint, but without visible stress fractures. I find her unflappability maddening, because in most respects, especially now when my rages materialize and disappear with the intensity of summer thunderstorms, I take after my father. I'm high-strung and bitchy, and I know she judges me for it, wonders if I'll ever have a stable family, a stable life. But something in me knows that in time, in a decade, when I'll probably be ready to have a baby, my mother's temperament will kick in. It's not that I hope to be like her; it's that I know I will be.

I'm twenty-two, living in Brooklyn with my best friend from college, thank God, since my night fears have returned with a vengeance fueled by local New York news. We live in a safe neighborhood littered with more strollers than rapists, but the prewar building creaks horribly in the middle of the night, and I hate coming home late when Allison is already asleep and I have to greet the dark apartment alone. It wasn't this way in college. College was safe, easy. I made friends and drank a lot and got into critical theory and developed a strong attachment to my English professor's six-year-old daughter, whom I picked up from school every day. I hadn't had a single steady boyfriend, but in my senior year, I fell in love with a man I'd decided I'd like to marry. Our babies would be redheads. But he wasn't my boyfriend. He was a grad student with whom I spent every waking minute. He prized his sexual freedoms and kept our physical encounters brief and illicit.

When I move to New York, I am still in love with him. I'm thinking about him one night when I go for a drink with a friend I've met on a film set, where I work as an assistant to a pretty famous actor. It's a great job to talk about at cocktail parties, except that I don't go to cocktail parties. This actor is sixty years old, and my whole life since I've moved to the city has been spent with old men, some of them very famous. It's not at all what my college professors had told me New York would be like. They'd assured me that city boys, so different from the Midwesterners who were taken aback by my brassiness, would be knocking down my door. But I haven't met a man under sixty in the six months I've lived here. When I see my friend, she's near tears. Her doctor has just told her that she has precancerous cells on her cervix, and that she may have only a few years to have children. She is twenty-three, single, and broke. I am horrified by her dilemma, but a strange pleasure comes over me as I talk to her about it. I feel deliciously adult, as if I am trying on my mother's dresses, as I talk to her about fertility and frozen embryos. These are words I've never spoken before, but they're rolling richly around in my mouth. I feel gloriously grown up advising her to just damn it all and get pregnant. She has a supportive family in the city, and if this is important to her and she loses her chance to have babies, she will regret it for the rest of her life. It's what I would do if I were you, I tell her.

I hold my legs in the air, my gynecologist's gloved hand deep inside me. She's fiddling with my cervix while feeling my belly with her other hand, looking into the middle distance with her head cocked, as if she were tuning a piano. I'm twenty-six years old, and it's taken the better part of four years in this city to find a gynecologist who didn't creep me out. When I finally settled on this one, it wasn't tough to deduce that it was because she reminded me of my mother—soft and gray and affirming, and so calm about everything. It was good timing, since I'm now about a year into a rocky, intense relationship with a guy I met at work. Since I found her, this warm woman has micromanaged my birth control (I can't have estrogen because of a clotting factor) and done helpful things like send me on vacation with plenty of UTI antibiotics, just in case.

"They don't seem to have grown any," she now says, in reference to my uterine fibroid tumors. They're not a big deal, these fibroids. I've known about them for a while, and something like 30 percent of women have them. The thing about mine is that at my last visit, my doctor had noticed that they'd started to grow, and their placement in my body means that they could potentially obstruct an embryo from attaching to the uterine wall. I hadn't panicked six months ago, when she'd told me. I'd called my mom, who looked it all up for me and said that I needed to exercise more and eat vegetables to keep them from growing. I'd done just that, and lost a lot of weight, so my doctor's assertion that these unwanted growths had not claimed new territory was an expected relief. Pop. Her glove is off, she's smiling at me. "So everything's fine," I say calmly, to double-check. "Oh yeah, you're great," she replies, chuckling easily. "I just wish you'd hurry up and get married and have kids in the next couple years, so we don't have to worry anymore." Her remark hits me like a skillet in the face. I cannot see or hear anything. I'm twenty-six years old.

Fifteen minutes later, I'm on my cell phone, careering wildly back to my office, jabbering incoherently to my friend Deborah. She's soothing me, saying, "Just tell Andrew what the doctor said. Tell him you want to get married and have babies. He'll do it. He loves you." You're right, I say, knowing full well that I will never say this. Andrew doesn't actually love me. And I don't actually love him. And on some level, I know this. I know it well enough to know that I would never dream of proposing marriage, or even telling him the truth about why I'm awake in bed at night, petrified that every twinge in my own abdomen is an ounce of my fertility evaporating, another minute in those "next couple of years" ticking by. For the first time in my life, in addition to my fears about finding love, I am scared of finding motherhood. And I understand that the two are linked, and I understand that the man I'm with is not the one who is going to give me babies. But I don't want to be alone at night, and so I do not leave him today. I will never leave him, in fact. I will wait until he leaves me—again and again, until it finally takes. It is the first of my conscious gambles. I'm betting that the system—the ride of life that will make me a mother and a wife, the assurances of my friends and family, my confidence in fate and romance—will win out. Betting that I cannot screw up this recipe: I am guaranteed a baby-cake, a family-pie, at the end of it, even if I leave out the baking soda, even if I waste another two years with a guy who's not right for me, even if I stop monitoring the very condition that has me so scared. Because the other thing I know, walking back to work from the office of my laid-back doctor who says I can have babies if I act fast, is that the only action I'm going to take is not about leaving Andrew. My choice is to leave her. I will not see her again, ever. After today, I will not see any gynecologist again for three years.

I am twenty-eight years old, and at a Bruce Springsteen concert at

Shea Stadium with my thirty-nine-year-old friend Lisa. I've got a new job that I'll start in a few weeks, and I'm making grown-up money. I've moved into my own apartment. And I am joyously, gloriously over Andrew. The weight of romantic and financial anxiety is off my heart for the first time I can remember, and I feel buoyant, in love with being single. I've recently been to a party where a palm reader told my fortune. She said that I would fall in love a year from now, and that I would have two sons, and, if I want one, a daughter. She also had me pick a tarot card. She told me that the one I'd chosen meant that I should stop looking for love and family; it will find me. This is just what I wanted to hear. It's not that I believe the palm reader, exactly. But this message is what I have always believed, or wanted to believe, what everyone has always told me. It will just happen, it's not about me making choices to ensure it. Lisa tells me that she has also just been to a psychic, who has told her that this year will be good for a baby. I laugh a little, thinking that this is what it's come to. We're smart women who make daily choices about what we do, where we go, what we eat, how we earn and spend our money. But our solace about our future as mothers comes from dim psychic predictions made by strangers.

But even as I smile, a powerful anxiety curls in my belly. Lisa's been trying for several years. I fear that she's not going to have her baby. She's always talking about it, planning on it, but it's not happening. I know she has the same fear. She advises me to freeze my eggs. She says if she'd frozen hers before she was thirty, she would have been pregnant by now. "I know it sounds silly," she tells me every time we have this talk and I wrinkle my nose. "But think of yourself at my age, knowing that you could have done this for yourself and didn't." This conversation always makes me feel nauseous. Lisa knows about my fibroids, knows that seeds of doubt have already sprouted in me and that they are growing, whether or not my long-ignored tumors are.

Lisa is the first person in my life who doesn't feed me the same line every parent and friend and tour guide and brochure and psychic has. She never tells me that everything is going to fall into place around the next bend. Every time we talk about freezing eggs I feel like she's telling me that she doesn't think that my family is just going to find me. I don't want to believe her. But on some level, I know that she is right where everyone else has been wrong. Nothing has ever been the way that people said it would be. What she's advocating is action. She's telling me to make a choice to be vigilant, to take responsibility for making sure things turn out the way I want them to. But I don't want to wrestle the monsters that are threatening me. I want to sleep through their creaky invasions and wake up to a house that's full of children and animals and a husband. And I know on that cool night that that's exactly what I'm going to do. I'm not going to take matters into my own hands. I am going to gamble again, wager everything that my partner, my babies, my home will indeed appear around the next turn, maybe at my new job, maybe in my new neighborhood.

I'm sitting on my stoop, finishing a pack of cigarettes. It's almost a year since Lisa and I went to see Bruce Springsteen. A few days after the concert, she told me that she was three months pregnant. She gave birth to her healthy son in April, three months after her fortieth birthday. I love my job, and have sped ahead this year like never before. I have loved my hiatus from heartbreak, played with my friends, have become jubilantly self-sufficient in my own apartment—fixing bookcases, tricking up my computer and stereo, reading whatever I want to until whatever time I want to. But a little of my euphoria over professional success and an independent life is seeping away, and it scares me. Because biological desire is starting to claw at my insides. This year, not just Lisa but three other friends have had babies. I've been put off enough by the immersion in episiotomies and breast pumps to know that I am not yet ready to give up my nights out, my mobility, my blooming careerism. But I've also held these new humans and felt at home, felt my desire for my own beating heavily against my heart. I've taken to mothering my younger friends and colleagues, clucking and offering steadying, sage advice. At work, I remain high-strung and bitchy, but my mother's disposition, which I always knew lurked inside me, is pushing its way out even before my babies do.

It reaffirms the possibility that I have it in me to be a parent, a single parent, if need be. I often make this point—to myself and others—but in truth, I'm shaky on it. Having a baby on my own seems like a different ball game than it did six years ago, when I advised my friend with precancerous cells on her cervix to do it. Because as proud as I am of fixing my own bookcases, the hammer and nails stuff is actually pretty taxing, and doing it leaves me feeling lonely and overwhelmed by unsharable responsibility. I think about hauling a stroller in and out of subways by myself and wonder if I'm brave enough to sign up for that kind of solitary burden. And I couldn't have just one; couldn't yoke a child with an unsharable responsibility of her own. I think of hauling a *double* stroller. Shamefacedly, I now check the ages of the couples in the wedding section of the *New York Times*, thrilling when I see a woman who's thirty-four, thirty-five, thirty-nine.

I try never to think about my fibroids. I still haven't been back to a gynecologist, though every few months, I make up my mind to call a new one and set up an appointment. I make little deals with myself while I'm outside smoking, which I have read is bad for fertility. If I go in for the night before I have this last one, everything will be fine. It's a lot of work to actually quit, especially when it's one of my few guilty pleasures; it's been a long time since I've had sex. Besides, women have it easy when it comes to quitting. When they get pregnant, they don't even have to make a choice; it just happens.

This weekend, a friend threw a party and hired Glenda, the same palm reader who told my fortune months ago. I remembered my last fortune very precisely; I'd written it down in my date book. I was surprised and spooked by the fact that her forecast for me was almost exactly the same as it had been then, though I did not remind her of anything she'd told me before. She even used the same words while examining my hand and proclaiming in a squeaky voice, "You think a lot. You think too much." The only thing that was different was that I'd been downgraded to one son and one dubious-sounding daughter. "Have you had a miscarriage yet?" she asked me searchingly. No, I said. She looked away and said, "Well, your boy is free and clear." As for the mate she'd promised I'd be finding right about now, he's apparently on his way. "Everything is going to change for you next month," she said. And when I picked the tarot card, she told me that

I'd selected the one that instructed me to stop looking for my love and babies. "They will find you," she said.

I put out my last cigarette and go inside and climb into my queen-size bed, where I lie for a while in the silence that tells me that I am the last person awake in my house.

MAUD CASEY

The Rise from the Earth (So Far)

IN THE TREES, THE SCI-FI CHIRRING of the seventeen-year cicada's call to mate haunts the air. May 2004: war and now, but of course, pestilence. Here in Baltimore, where I've landed in my nomadic thirties, the cicadas lie spent on the sidewalk, fluttering one postcoital wing for hours before they die. With their buggy red eyes and cartoony paned wings, they careen lustily into car windshields, leaving behind a sticky smear of longing that no amount of window-wiper fluid can erase. One of them (precoital frenzy? a swoon of dying delirium?) flings itself into the arm of my windshield wiper as I drive down the road, and I leave its corpse stuck there for days, out of a kind of admiration. These fiercely determined insects seem symbolic, important. A symbol of what? Biological destiny? Perseverance? I imagine they have something to tell me, a thirty-five-year-old woman who may someday want to have children. After seventeen years of muffling their cicada desire deep in the dirt, these determined insects have risen from the earth on the same night with a sole purpose: to breed. All day long, that sci-fi chirr, a plea in the trees, where they've instinctively climbed from their underground burrows to molt and get it on.

The male cicadas vibrate their timbals, drumlike abdominal membranes. *Chirr, chirr.* Let's do this thing and make more of us to burrow deep and rise to do this thing again.

The seventeen-year cicadas are also known as magicadas and there is, in fact, something magical about their trilling refrain. It's not a question. In the face of the resolute nature of nature, my own quandary over whether or not to have children takes on a poignant human quality. No, the magicadas are unfettered and uninhibited in their commitment to their evolutionary task. They don't wonder whether they should bring their bug children into an overpopulated world of torture, natural disaster, poverty, disease, cruelty, injustice, not to mention the sometimes numbingly banal trudge, trudge of just getting through the days. They don't wonder how they'll afford it all or whether they'll be good parents because parenthood isn't part of the deal. For them, there is no why. Their mission is pure: rise up, breed, and die. I can't help feeling a slightly bizarre twinge of jealousy for the *chirr, chirr* in the trees outside. The female cicada will lay between two hundred and six hundred eggs that will hatch into cicada nymphs, who will burrow deep, suck the juices from trees in order to survive for seventeen more years. In the seventeenth year, the nymphs will climb the trees and emerge from their split skin to begin the *chirr, chirr* and find female cicadas with whom to mate. Burrow, rise up, breed, die, repeat. In the throes of my reproductive years, surrounded by friends who are new parents or on the verge of parenthood, the cicada's song has a keen resonance. Lurking behind the buzz in the trees, I hear a different sort of sweet biological chirping: Will you or won't you? Will you or won't you?

But for me, the question of *how* to be pregnant occurs simultaneously with the will I or won't I, the why and the whether. At the age of eighteen (around the same time the last batch of recently hatched magicadas had gone underground to nourish themselves with the juice of trees), I was diagnosed with bipolar disorder. Ever since, I've been taking medication to prevent a relapse into the whirling circus of psychosis and mania, where, among other things, my mother was disguised as Madonna sending me messages through her videos, and wearing shoes meant you walked on souls (soles). This medication also prevents another foray into the dead land of depression where the idea of my mother-cum-Madonna seemed like the good old days and putting on shoes took all of the energy I could muster, never mind worrying about trampling spirits. The bipolar disorder is located somewhere between my animal body and my human mind, a combination of genes that have ingeniously conspired with that various and complex thing called life to create a chemical imbalance, and I am down-on-my-knees grateful for the relief the medication brings me. I often find myself staring at the drugs in my cupped palm in amazement and wonder: Lexipro, an antidepressant, as a tiny white bitter-but-blissful tablet; the antidepressant Wellbutrin is a curvaceously round, pink cheerleader of a pill; and the Stepford wife–sounding mood stabilizer Tegretol is a humble, medium-size pill that looks deceptively like a run-of-the-mill aspirin. My marriage to this medical cocktail has been a relatively stable and happy one, and pregnancy would require a separation. The most alarming aspect of this separation is the break from Tegretol, that miracle aspirin that keeps my moods, well, stabilized.

According to various studies, if a woman takes Tegretol during the first trimester of pregnancy, there are risks: a 1 percent risk for neural tube defects, such as spina bifida (versus the .01 percent risk associated with women not taking Tegretol); an increased risk of heart malformations, cleft lip and stunted growth (particularly in head size); and a heightened risk of abnormally large spaces between the nose and upper lip. The research is incomplete and hard to come by, because there are ethical considerations—more risks—involved with doing research on pregnant women.

There is also evidence that the more depressive or manic episodes a woman has had, the more likely she is to have another. In other words, there's a cumulative effect, a building of one upon the other, like sedimentary layers. There have been studies in which over half of the women observed relapsed into the whirling circus when they went off their medication during the first trimester of pregnancy. Questions beget questions. Will I risk my hard-won mental health for motherhood? Certainly there will be scientific progress. There are National Institutes of Mental Health initiatives dedicated to finding mood stabilizers that are less dangerous during pregnancy. Women with bipolar disorder can, and do, work closely with psychiatrists and therapists during the first trimester to monitor mood shifts, to keep track, to balance the risk of relapse with the risk of birth defects. So, let's say that I will risk it. Let's say I begin to consider the why and the whether. Again, the answers are questions. What if my child, having been spared spina bifida, a deformed heart, or a tiny nose, inherits the worst version of my haunted genes and it turns out that he or she is shouldering the 10–15 percent risk of suicide that accompanies bipolar disorder? How do I assess that ephemeral feeling of mania, see it through the eyes of my child? The day I sat on my bed as a freshman in college and felt for the first time something alive rising in my throat, terrifying and exciting, slippery silver like mercury streaking up my throat, the elusive thing that later would be identified as a disease? This becomes something wholly different if that hard-to-pin-down feeling might be passed on to my child in some form. How do I measure the sublingual, elusive grind of depression, the days of staring at my bedroom ceiling, its blankness a generous projection of my mind? Because when I think of having a child, I fear the day I look into her eyes and realize the unquantifiable "it" of bipolar disorder is in her too. The day I realize that, knowing the risk involved, I did this to him. Will I risk the mental health of my unborn child?

Like the branches of bonsai trees, the questions loop back on themselves to the central trunk: why would I want to have a child? Before the questions, there is history, the history of imagined and accidental pregnancy. *Chirr, chirr,* a rattling of my grown lady timbals, a fugue in three parts.

1986: The Rise from the Earth, Part I

Wherein I mourn the blood that blossoms in the toilet bowl and refuse sanitary napkins from the nurse in the university infirmary though my underwear is soaked with blood. The nurse tells me she's not going to change the sheets until I agree to wear the sanitary napkins. I've made my bed and, well, I can lie in it. Her sentences are sharp, slicing through the air. In this clipped, unyielding voice, she speaks often about choices—to sleep or not to sleep, to fight the sleeping pills meant to sleep me through this part until the part when my parents arrive to take me home. I've had my first manic episode, though no one knows it yet. The consensus so far is that this unusual behavior may be the result of experimenting with drugs or brought on by lack of sleep, neither of which are unusual for a second-semester college freshman. There is an authority to the nurse's voice like a wall that my slippery-silver mercury mind leans against gratefully, but she doesn't understand, and I can't explain. I am pregnant and I am dying. I'm not sure if I say it out loud. It's a poetic declaration that I refuse to surrender. Until today, when the blood tells me that I'm miscarrying, that I'm losing the baby.

"It's your period," the nurse says. Clearly, I have said something out loud. She waits outside the open bathroom door, her back turned to me. Suicide watch precludes closed doors. "I'm not going to say it again," she says, though she just has and she will. Her smile is less friendly than her regular face, which is reassuringly and determinedly expressionless. I return to lie in the bed to bleed out my phantom baby. Outside the window, students walk along the trimmed paths, backpacks slung over their shoulders, on their way to class in the gauzy beginnings of the spring light that's come to swaddle the remnants of cold winter air. I bleed and wait for the public safety officer on night duty who, for the past several nights, has sat by my bed and talked about God, His goodness, His greatness. I don't think I believe in God, but the phantom baby is close to something bigger than myself, something spiritual. The public safety officer is a tall, thin black man with high cheekbones set in an angular face and an orangeish Afro that escapes the sides of his public safety hat as if it is flying away like the angels he speaks about. The night shift begins when everyone else goes home. When he returns home in the morning, his family—he has a wife and three children, he tells me—has already left for the regular daytime world. He lives an opposite life that I can relate to. Together, I imagine, he and I exist outside of time. We are transcendent, like the child my mind has conceived on the heels of first love. "God loves you," the public safety officer says, and when I tell him that I'm losing my baby, tears run over the sharp bones in his face and he holds my hand, laying long, spindly fingers across my own short ones, soon to be fat from medication. "I'll pray for you and your lost child," he says.

He knows there is no lost child. The nurse told him in one concise, chiseled sentence as she headed out the door, explaining the thick pad placed between my body and the mattress. But he was generous, willing to respect my hallucinated pregnancy, this excess of feeling. Beyond the hot-blood zing of mania, the rush of psychosis to make signs and symbols out of every dull scrap of life, did the excess of feeling have something to do with a desire to create life? Or did it have more to do with a desire to grow up? Something in me wanted to go beyond myself the way this man with real faith seemed able to, crying for me, a crazy girl yearning wildly, recklessly, for a tangible tragedy, blood she could see. At eighteen, when I couldn't see into my future, when I couldn't see much of anything except myself, the possibility of having a child, the idea of it, felt powerful.

1998: The Rise from the Earth, Part II

Wherein the nurses at the psychiatric hospital tell my family I've been restrained because I smeared menstrual blood on the walls of my room. Several weeks into my second hospitalization for bipolar disorder, I've been restrained several times. When this happens, my arms and legs are belted to a gurney so that I can't move in any direction. I have a memory of being upside down as mental-health workers secured my arms and my legs, but I don't think that really happened. I don't remember smearing my menstrual blood on the walls of my room, either. What I do remember is that I believed I was miscarrying again. I bled, and years of anger and sadness were focused to a sharp point of grief that became the loss of the second phantom child I believed I was carrying. Again, an excess of feeling, a bursting of love in my head bursting for so many things, but the biggest explosion, the one that seemed the most palpable was for this lost child. (Years from my first love, the father of this child depended on the direction of my nostalgic hyperbole on any given day—yes, yes, that one-night stand during that year of temping and too much drinking, he's the one.)

Who was this imaginary kid? The pattern of delusional pregnancies are rich material for my therapist, but beyond that, and beyond assigning importance to random men, or perhaps behind, or next to these things, there was something about the idea of being pregnant, even in my psychotic fantasies, that was empowering. It was mine, *more* than mine. In a funny way, it felt as though I were escaping my body, transcending it. But there was no escape, except for the blood on the walls translated into a child, and maybe my body, its blood coursing at that point for over ten years with various life-saving medications—psychotropic drugs, SSRIs, mood stabilizers—had a hunch that down the line, when the question of real children came up, so, too, would the question of the inheritance of these spooky genes. You might not be able to do this pregnancy thing, my coursing blood might have been saying, rehearsing fearfully. Prepare yourself. Practice this loss.

2001: The Rise from the Earth, Part III

Wherein Tegretol reduces the efficacy of the birth control pill by 3 percent. This information is provided in the Drug Interaction section of the pharmacy pamphlet that accompanies Tegretol prescriptions. It is also provided in the Drug Interaction section of the pharmacy pamphlet that accompanies birth control prescriptions. In 2001, I'd refilled my Tegretol prescription approximately forty-eight times (Tegretol was fairly recent, after ten years of lithium and a month of Neurontin) and my birth control pill approximately eighteen times. I never once read these pamphlets. Over the years, I'd grown accustomed to leaning on the hard wall of authoritative voices like the nurse's in the college infirmary. I leaned willingly, eyes closed. I'd been taking medication for fifteen years, and I never investigated my own illness, grew suspiciously sleepy the second paragraph into the books dedicated to exploring the mystery of bipolar disorder, never researched the medications that ran through my blood. I assumed that hard wall of voices, the *they* (who were "they" exactly?) to whom I had gratefully turned over the reins, would intuit and intervene if there were a problem, and no one—not my gynecologist, not my primary care physician, not my psychiatrist, not my therapist—mentioned that Tegretol decreased the effectiveness of birth control pills, and I never asked. It didn't occur to me. At thirty-two, an age when many of the women I knew were trying to get pregnant, some turning to fertility treatments, I found myself accidentally pregnant.

When the pink line appeared in the tiny window of the EPT, I didn't experience the clarity of vision, the rush of power and certainty, that feeling of transcendence, that came with the phantom babies. Instead, it was as if I had been visiting this EconoLodge body of mine and now wanted to sneak out, having wrecked the motel room. The choice was clear, the only sensible thing to do, my boyfriend of two years and I agreed. At the time, I lived in an eleven-by-fourteen studio in Fort Greene, Brooklyn. There was my financial insecurity to consider, his financial insecurity, and our inability to negotiate our relationship, but the bottom line, that relentless cement foundation against which I kept knocking my head, was my medication. "You can't go off your medication," my boyfriend said, and for the millionth time, I agreed with this assessment, resting my weary, guilty head against that familiar wall of hard, official sound.

My period was irregular with the birth control pills. It wasn't unusual that I would skip a period or two. But when I took the EPT test, I was thirteen weeks pregnant, as pregnant as several friends who had recently called to tell me they were far enough along to feel safe letting people know. (Again, no dearth of material for my therapist.) I wanted my boyfriend to love me more than he did, and maybe there was a kamikaze way in which I wanted to be pregnant. But there was more to it than this. Why had I ignored my body for so long? This is the part of the fugue that is the most difficult to sing, because my voice quavers, more of a squeak than a song: I was willfully estranged from my own body, and I had been for years. I wasn't so much married to my medication as I was its child, as eager at thirty-two as I was at eighteen to escape my corporeal self.

I was far enough along that I received general anesthesia when I had the abortion. I didn't think about my uneasy relationship to my body, or birth defects, or Godlike power. I didn't think anything at all. There was no excess of feeling. There was me and four teenage girls changing in a small room behind screens into hospital gowns; there was waiting side by side in orange plastic chairs until our names were called; there was tinny Top 40 music playing on a radio in the room as I counted backward; there was deep, dark anesthetized sleep; and there was waking up in a different room with the same four girls, one of whom was vomiting into a bedpan.

My gynecologist did a sonogram before she sent me to the clinic (because she was not allowed by the hospital that employed her to perform abortions), and, in a move that still puzzles me, she printed it out and gave me a copy. It was then, with the generic, grainy moonscape snapshot in my hand that I thought of neural tube defects, heart malformations, oral cleft and urinary tract deformities, facial abnormalities. At the center of the moonscape sonogram was the fluid comma, the beginnings of shadow limbs, a shadow head. I've never doubted I did the right thing, but I carried that grainy photograph in my purse for days. I was horrified that this shadow being might have been deformed by the very medication on which I relied, that I'd subjected it to this possibility for thirteen weeks. When I was eighteen, and again at twenty-nine, imagining pregnancy was part of an effort to jump ship, but after thirteen weeks of the real thing, I realized that having a child would require that I stop trying to throw myself over the railing, that I stay on board. It would require that I really inhabit this body of mine, but not only did my body seem more real than it ever had before, my diagnosis did, too. I wouldn't be able to lean against the wall of voices, because if I became a mother, "they" would become me.

Soon after the abortion, my boyfriend told me he didn't want to have children. I pulled the sonogram out of the file in my office where I'd tucked it between papers. I stared again at the comma curve, its cells halted in the course of their dividing and dividing, and thought, this is it. This is as close to motherhood as I'll ever get. A friend of mine, a mother for eight years, said to me recently that she'd gone to see *The Addams Family* with her daughter and that something in the characters' celebration of doom and gloom made her nostalgic for a quality in herself that she considers prematernal: self-destruction. She didn't mean that there aren't self-destructive mothers, or that motherhood was a magical land of self-love and rose-smelling diapers. She was talking about something else, a rise from the earth, an occasional distance from your self. A distance that, say, might encourage you to not spend an inordinate amount of time staring at the sonogram of an aborted fetus.

The sonogram is gone, and the boyfriend is, too. These days, I'm free to imagine the levitation involved in motherhood. I think of the book of poetry by Wislawa Szymborska that I marked up with Magic Marker the second time I was in the hospital. The markings are wild and cryptic, childlike. The ink soaks through the pages where certain lines are underlined two or three times. One of my favorite Szymborska poems, "Grace," is free of my frantic graffiti. The last stanza is a different kind of search for meaning. It's an appreciation for the miracle of someone else's survival. *So you're here?/Straight from a moment ajar?/The net had one eyehole, and you got through it?/There's no end to my wonder, my silence./Listen how fast your heart beats in me.* If I figure out the how and the why and the whether, and someday I have a child, and somewhere between her animal body and her human mind a combination of genes conspire with life to create silver mercury that rises in her throat, and then steals that mercury so that there are only blank walls, I want to remember this poem so that I can offer it to her.

Chirr, chirr. And so on and so on and so on. The cycle is so simple and pleasing, so foreign to this human in her animal body.

STEPHANIE GRANT

Beyond Biology

WHEN MY MOTHER WAS ALIVE, people frequently commented on my resemblance to her. I have my father's coloring but my mother's facial structure, the shape and set of her eyes and mouth. Less obviously, I also have her hands: swollen joints and crooked fingers, age spots and prominent veins, what she called "crepey" skin. Once, a few months after she died, I found myself waking from an uneasy sleep. I was at a writers' colony, in upstate New York, curled in an unfamiliar bed. I had slept on my side, with one arm supporting my head, the second resting on the mattress before me, my hand inches from my face. As I blinked awake, I saw the hand and started: it was my mother's. I thought she was still alive, kneeling beside my bed, her hand about to touch my face. When I remembered she was dead, I still felt her presence in the room. Not a ghostly presence, but rather a sensation of—how to put it?—containing her. I became conscious of myself as a vessel for her presence in the world. Lying still in bed, afraid of breaking this thread, afraid of dispelling this newfound awareness, I recalled her funeral and how her friends had approached me, arms open, hands raised, their aging faces grief-struck and expectant: they wanted a last touch, a last embrace with my mother.

I became a mother for the first time at age forty, when my partner, Ara, gave birth to twin girls. We started trying to conceive when we were both thirty-seven, which is to say we began doing at-home, intracervical inseminations. A good friend who is a nurse practitioner lent us a speculum and taught me how to locate Ara's cervix. The first insemination took place on the turn of the millennium, New Year's Eve 1999, and then again, New Year's Day 2000. We did it at a friend's house in Provincetown, with several more friends present. The timing felt auspicious, and our mood was celebratory, almost ecstatic: what better way to begin the next century? Of course, we expected to get pregnant right away.

But after a year spent trying, we finally went to a fertility clinic and exchanged our do-it-yourself, at-home inseminations for in-office, intrauterine inseminations. At a night class peopled with unhappy, infertile couples, I learned how to give Ara shots in the belly, of a hormone that encouraged egg follicle development, as well as shots in the butt, of a hormone that induced ovulation. Five sullen men and I practiced giving shots to an orange. When, a few months later, the test strip—at last—turned purplish-blue, we raced to the fertility clinic to look for the beating pulse on the monitor. There were two.

Two! We were thrilled and overwhelmed and slightly terrified. I had seen formidable, capable women almost incapacitated by the arrival of a single child. How on earth would we manage two? Still, the news felt like a gift, like compensation for all that time spent trying. Ara and I exchanged a stupefied glance: we were going to be the parents of twins? As I groped around the tiny, white-and-chrome examining room for a supportive chair, I became aware of another, less-welcome feeling asserting itself. Not regret, exactly, but a kind of panic at the realization that now I might not get to have a baby myself. It had been easy, two years before, to agree to Ara's conceiving first; it had been less easy to wait patiently month after month when she did not. Recently, I'd had a battery of tests to assess my fertility; soon, we'd decided, if Ara did not get pregnant, I would begin to try.

In the doctor's office, in the midst of all that joy, I was embarrassed by my ignoble feelings. *What about me?* I wanted to shout, selfishly, gracelessly. *What about my pregnancy? What about my baby?*

One of the reasons we started trying to have children as late as we did was that I had a lot of ambivalence about becoming a mother. My own mother loved her children intensely, but, like many women of her generation, seemed to lose her sense of an independent self, of any identity beyond mother. And with that went much of her capacity for joy. I can still picture her sitting at our kitchen table, hunched over the local paper, reading in a kind of fugue state that blocked out all of our existences. In my memory, she is slowly eating those bright-orange cheese crackers that come in cellophane packaging, placing them one by one on her tongue. Her pleasures had gotten so small.

For most of my adult life, I have been afraid of being as overwhelmed and unhappy as she was in motherhood, afraid of forfeiting the things that make my life rich and satisfying, and so very different from hers, namely my intellectual and creative work. Since my daughters' arrival, I have worked hard to maintain my writing life. It hasn't been easy, but it hasn't been impossible, either. Now that I have two children, it is the third, I imagine, that will bring about the forfeitures I've so dreaded. My resources—physical, financial, emotional—are not limitless. I'm afraid another child will precipitate the end of self, the end of literary ambition.

In my twenties, I made a promise to myself to publish two novels before having a child. I wanted my identity as a writer to be solid, undeniable, *irreversible* before parenthood. A month before the girls were born, I sent a completed draft of my second novel to my agent, but he did not agree that the book was done. So here I am, at forty-two, already having broken a promise to myself, working feverishly to finish my sophomore novel. Given my not-inconsiderable ambition, how can I even contemplate a third child?

Because—it almost chagrins me to write—I ache to conceive and give birth. In fact, the intensity and persistence of the desire surprises me. Certainly it contradicts my deeply held belief that biology does not, in and of itself, make family. What is that desire, I ask myself, now that I am already a parent? Now that my daughters, Augusta and

Josephine, have become spectacular realities? It is not that I want more from them, or that I experience our connection as somehow deficient because we are not biologically related. When Ara was breast-feeding, I used to joke that she was the girls' wet nurse, so keenly did I feel my legitimacy, my primacy even, as their mother. My desire to conceive is not about my daughters then, but about me. About what I get to do and have in this life.

Perhaps it is the experience that I want: conceiving, carrying, giving birth to a child, the singularity of biological motherhood? It's true that all of these have intense, unparalleled allure. I've attended a total of four births now, including those of my daughters, and each was transcendent. Of course, I would like to know that particular experience intimately. But, the truth is, I've seen too much to romanticize the birth process entirely. I've witnessed miserable pregnancies in which there was no reprieve from morning sickness. And I know the physical costs of delivery—in tearing, in (temporary) neuropathy, in incontinence—to the biological mother. As I've gotten older, the allure of the experience has faded considerably. Now I yearn for the baby, not the pregnancy.

Some days my desire feels like a test of character: why can't I simply accept the extraordinary gift of my daughters and let go of what I think I want but don't have? On these days, my longing for a biological child feels frivolous and narcissistic. It is merely the wish for continuity, for the shock and pleasure of heredity, the passing down of my own physical traits, and those of my family.

Surrounding the girls is a perpetual susurrus of conversation about just whom, in Ara's family, they resemble. That I stand forever outside these conversations is a source of pain and disappointment. How could it not be? As the girls grow and change, the conversation also changes: now Augusta looks less like Ara and more like Ara's father; now we can see the resemblance between Josephine and her first cousins. I did not anticipate the sting of these conversations before the girls were born, because I did not—could not—anticipate the intensity of motherhood; because I did not yet know the immense pleasure my connection to the girls would bring me. But the desire to become a biological mother is about more than the pleasure of heredity and cannot, I think, be reduced to it. The desire is bigger and more irrational than that.

When I think of not having a child, I think first of what I'm denying myself. But then I think of my mother and what I'm denying her. It is as if we had made an explicit agreement that I would have biological children, and that this agreement was the true stuff, the sticky marrow, of our relationship: my willingness to repeat her choice. Ultimately, when I imagine not conceiving, I think of what I'm denying the two of us, together: the pleasure of "reproduction." My mother expressed through me. Not in my return to my mother's womb, but in her reemergence through mine.

Several years before the girls were born, my mother died of ovarian cancer, at home, with the family present. During her last days, I remember thinking that what she really wanted was not any of the things she asked us for—ice chips, and lime-green Popsicles, and liquid drops of morphine. What she wanted was for us to keep her alive. Having a biological child seems a way of doing that, a means of recuperating her physical self in the world. How can I relinquish that?

And yet, from time to time, the girls *do* remind me of members of my family. Augusta is tall for her age and deliciously plump. Like my mother, she is quite affectionate. From the moment she first had control of her head, she began nuzzling people's necks. At day care, she is known as an inveterate hugger who does not always secure consent in advance of her embraces. At home, she hurls herself at Josephine, or at Ara and me, with full-body abandon. When Augusta tackles me, I can feel my mother's fierce love let loose in the room.

Josephine is considerably smaller than Augusta, undersize, but with a prominent belly that she presents proudly with each step. When she runs, she gets up on her toes, arms raised at her sides in anticipation of something good; she is eager with her smiles. Some days, I can see my great-aunt Grace Reilly (my mother's mother's sister) in Josephine, whose mischievous smile recalls Grace's open face. My mother took care of Grace when she was elderly, and over time they came to resemble one another. If Josephine were my biological daughter, I'm certain my family would remark on how much of a Reilly she looks. It is tempting to trace a line of continuity from

Grace, down to my mother, through me, and, finally, to Josephine.

The more I think about heredity, the more slippery it gets. I cannot count the number of people who've told me that Josephine resembles me—people who know I'm not her biological mother. It's true that Ara and I chose a sperm donor precisely to encourage resemblance (to minimize my vulnerability as the non-biological mother), but it is interesting to hear the ways in which people see or experience our physical relatedness. Many insist on Josephine's likeness to me, and Augusta's likeness to Ara—a neat, even satisfying, division. When this happens, heredity—physical resemblance—feels more like a story we tell ourselves than a factual accounting of traits. If heredity is more narrative than substance, I can't help but wonder, what are these people trying to say? What, beyond biology, is being communicated?

My father came to visit when the girls were about five months old. He had asked to come earlier, but I'd put him off. I'd wanted Ara and me to get a handle on caring for twins, establish ourselves as knowing what we were doing, before we put our mothering on display. But I also feared his response to his non-biological grandchildren. My dad is Irish Catholic and socially conservative; it took him many years to accept the fact that I was a lesbian. I'd trained myself to have very low expectations about how he received my world.

We have pictures from that visit. My father sits in a fraying upholstered rocker in our living room, Augusta bouncing and smiling on his knee, Josephine stacking cups at his feet. My father—big-boned and red-faced, with white, white hair (think Tip O'Neill)—looks ecstatic. Just minutes into his visit, he announced his delight with the girls. "Stepha," he told me, "they're beautiful. They look just like me!"

In fact, he said the girls resembled him so many times during his visit that I wondered if he had somehow imagined our reproductive technology as more advanced than it was. Then I realized what he was doing. With this rhetorical gesture—"They look just like me!"—my father was telling me that he recognized Augusta and Josephine as my children absolutely. He was telling me that we all resemble each other, that we all recognize each other, as family. The pleasure of heredity is the pleasure of recognition, is it not? *I see myself in you.*

Narcissistic recognition, but also existential recognition: *I see you seeing me. We see each other. I am being seen.*

When friends and acquaintances assign the likeness of one twin to Ara, and the likeness of the other to me, I think they, too, are recognizing my legitimacy. They are trying to make the lottery of conception—the surprise of Ara's conceiving two—the terrible unfairness of that (unfair, that is, to me)—they are trying to make it right, make it reasonable, make it just. There is a pleasing symmetry to our family, two mommies, two babies; how much more pleasing—and less challenging—if we were each the mother of one.

I'm supposed to be leading somewhere definite, to a conclusion in which I announce what I will do. But I still haven't decided. It occurs to me to do what is best for Josephine and Augusta and to ignore my own desires. But that is an old and familiar trap, my mother's trap, which I want to avoid.

Some days, it seems as though my urge to have a child is simply my wish that Augusta and Josephine once resided inside me, that I had contained them as my mother contained me. When Josephine locates her belly, as she often does with delight, when she looks up at me and smiles—she has a wide and beautiful mouth, her smile filled with intense and easy joy—when she points to her belly button and laughs, I am astonished that it does not reveal the literal truth of our connection. I am sad, too, that it does not uncover the invisible umbilicus between us, but must stand, instead, as a metaphor.

Of course, I believe in the power of metaphor, I am a writer after all. But I want more than metaphor; I want skin and flesh and bone. Like many writers, I have a literal cast of mind; my metaphors tend toward the actual, toward the concrete. Avoid abstractions, I tell my students, your metaphors should help us see as well as feel. Maybe that is the power of heredity: how it concretizes, how it locates grand and inchoate desires—the desire for immortality, say; or the desire for recognition and reproduction; the desire to see and be seen for who we are—how it locates these abstractions in the beautiful, literal bodies of our beautiful, literal children.

It is ironic, of course, but it is probably the writer in me who most keenly craves a biological child, even as this desire may contradict, or threaten, other writerly needs, like the need for time and solitude. It is the writer in me who needs to know what a particular thing *is,* rather than simply what it is *like.* And it is the writer in me who honors the irrational, even as I search out rational means of expressing it.

In other words, although I understand heredity to be partly a fiction—something we construct together—it's a fiction I'm eager to know.

Perhaps this is what I will pass on to my daughters: a vision of me sitting at the kitchen table, hunched over my computer, holding two contradictory ideas in my head, savoring them, side by side, on my tongue.

LARRY SMITH

The Daddy Dilemma

WAS LIFE SO PERFECT BEFORE THE MARGARITA INCIDENT? Sometimes I think it was. It was a life less examined at least. And that can be a good thing. The Margarita Incident involved—as those moments in life that somehow mean a lot often do—tequila. And a small child. And my fiancée for the past eight years, Piper.

Piper and I were having a particularly good time trading funny faces with a super-cute two-year-old in a Mexican restaurant in the East Village. We were riding what was up to that point the perfect buzz available to two people with dual incomes, decent rent, and no need to be home by ten p.m. to pay a babysitter, when she looked at me and asked: "You're not going to turn forty-two, freak out, and leave me for some twenty-seven-year-old eager to be a mom, are you?"

Good question. If I did turn forty-two and despaired about not being a dad, a logical solution would be to find a younger woman who wanted kids—there always seem to be a lot of them around. But I'd prefer to avoid that situation. Piper is the love of my life. She took long enough to find. I'd prefer to live out my life with her.

Here's the deal: we don't know if we want to have children. I'm about 65 percent for procreation; she's about 70 percent against. As we both slip into our mid-thirties, my own personal daddy dilemma has quietly taken on an urgency that I frankly didn't expect. I know that if I'm cutting out of work early to go to a soccer game, I really don't want to be the oldest guy passing out the orange slices, or, worse yet, have my ass kicked by some young dad during a bout of sideline rage. We don't need to breed tomorrow, but we can't wait another eight years, either. I think in decades, she thinks in days, and now, more so than she, I think it's time to figure this stuff out.

You know that illustration with a stylish woman talking on the phone, saying, "Oh my God, I forgot to have children"? I don't want to look up in 2015 and realize I'm a version of that woman (albeit one wearing a worn Philadelphia Eagles sweatshirt). That noise I'm beginning to hear is the sound of my sociological clock ticking.

From as early as I can remember, I've been told I would make a good father. My mother repeated this mantra often. For her it was a statement of fact, backed up by the care and interest I took in my little sister (as a four-year-old, I would declare, "Don't you drop that baby on the floor" whenever someone held her; this delighted my mom and made my dad quite nervous). If she and my father have raised me right, the idea of starting a family should be an attractive option. They've done their part to continue the great cycle of life. The pattern continues with me. So it was decreed. Or at least presumed. That was the plan. The Smith family name would continue.

We may have the most common surname in the United States, but our stock is special. The first Smith was my grandfather Morris, who died three years ago after an excellent run that began in the tiny Russian town of Minsk and ended ninety-one years later in a tony suburb of Philadelphia. Upon arrival, his family was anointed "Smith," a loose translation of the family name of Blacksmith, an irony not lost on the generations of Smith family men more comfortable at the racetrack than the metal shop. Morris became known to one and all as Smitty, a nickname you don't hear nearly enough anymore.

Smitty had two boys—my dad, Louis, and his brother, Uncle Ralph. Lou had two girls and me. Ralph got married to Kathy, with whom he shares a blissful, travel-full, kid-free existence. This makes me the only living male in my family still realistically likely to father children. "It's up to you," the first Smith said wistfully ten years ago at my sister's wedding, as my then-girlfriend looked on in horror.

But a funny thing happened on the way to the birthing class: I started dating a woman who could imagine a life for herself that did not involve children. We were still young, and the pregnancy craze among our friends was still years away, but the germ of an idea was planted. Maybe I didn't need to spread my seed. Taking a pass on parenthood was an option. Who knew?

She knew.

I met Piper in a dive breakfast joint in San Francisco nearly a decade ago. She was a girl who was used to being chased and good at rarely being caught. I eventually tricked her into dating me and have held on tight ever since. We've moved from the West Coast to Boston and finally New York, surviving deaths in the family, layoffs, landlords, legal troubles, terrorism, and all the other things that if they don't kill you make you stronger. Now, eight years into this thing, we've found ourselves with good jobs, great friends, and ten thousand songs on our iPod. The only things to keep alive are a couple of plants and a couple of cats. It's the perfect bohemian yuppie existence. Why mess with a good thing?

For an I-don't-want-to-grow-up guy in his twenties—which is to say, myself and most everyone I knew when I met her—Piper was a dream girl. The "Vows" column of the *New York Times* might say: "A child actress who grew up in Brookline, Mass., friends say Ms. Kerman exhibits the same comfortable ease munching on chicken wings and watching the game with the guys as she does preparing complicated Indian meals and discussing the latest article in the *Atlantic Monthly* with her Seven Sisters college alums. She loves wide-open spaces and horseback riding, yet has performed decorating miracles in her tiny New York City apartment. She's as at home in an organic garden in Berkeley as she is at a sample sale at Barney's. She has never considered diamonds to be a girl's best friend and was shocked when Mr. Smith presented her with seven gold rings and asked her to be his life partner. Although her many years of babysitting have afforded her a winning way with children (not to mention the ability to change a diaper with one hand in eight seconds), what Ms. Kerman really hopes for, she says, 'is a puppy.'"

Good deal, right? When all your fiancée demands is a baby bulldog, your buddies declare her Woman of the Year. What more could a dude want than a partner who isn't pressuring him to get married and make babies?

Many of my pals talk a good game about how their wives finally put the clamp down and demanded they do the deed, but deep down they're all glad the women forced the issue. They knew the day was coming; most of us want it to come, even if few of us ever feel the time is exactly right. No man really believes that he will ever be prepared to take another's life in his hands, a time described by one friend as "when all hell breaks loose."

Do I want all hell to break loose? Do I want, as another expecting father I know predicts, "life as we know it to end"? Like I said, about 65 percent of me thinks so. Besides all the good press, it's a big life experience, arguably the biggest. You don't go to Disney World without riding Space Mountain, right? But here's the rub: I don't *know* if I want to have kids—and yet it's up to me to convince my free-wheeling, sassy better half that it's what she wants as well. Ask around—I have—despite all the progress we've made in rejiggering gender roles, Piper and I are a rare breed.

Which leaves the ball bouncing perilously, nervously in my court. Can I know for sure? And if I can, how? When? I'm much more happy than not with the life I've created for myself, but I'm far from content. Everything about my thirty-six years on earth has pointed to career being the source of salvation, so I keep thinking I'll be satisfied when that's at the place I want it to be. But will it ever be? Won't the bar keep being raised? Isn't that what careerists do? Would a child make me see what's really important? You're a Cuervo Gold–slugging ass, lost and lonely, or worse . . . until a child enters your life. Maybe. In all honesty, I can't say for sure that a child would do more for my contentment quotient than any number of professional goals. Hell, anyone can make a baby. But only I can bring a really original new magazine into the world.

There's simply no way to know for sure. I see now that it would be easier if, like my other friends, I didn't have any choice. Guys don't have the same biological urges as women. Their partners get pregnant. They get scared. They get a grip. They turn to baby-loving mush. Digital photos ensue.

To Smith or not to Smith? Seemingly the oldest question in the book, and yet completely new territory for a Smith like me. My grandfather and father didn't even think to ask these questions. Their lives had expected plans and paths. Not so long ago, if you didn't have children you were possibly gay, probably single, perhaps infertile. My grandfather and father's American dream was the same: a better life for the next generation. That there would even be a next generation wasn't a question.

I picked Piper not because she's a fertile vessel and a future supermom, but because there is no one else I've ever met whom I want to be around so much of the time. She's a strong woman who looks hot with a tool belt and is by far the best candidate in our home to figure out the re-fi (not to mention the Wi-Fi). Finding a partner in this modern world is no small matter. Okay, so it took a little while—we got engaged after seven years—but I can say without further hesitation or equivocation that this is the woman for me.

Much to the astonishment and admiration of my pals who have felt the heat to get a family started, until recently, Piper and I actually hadn't spent much time talking about what I—and my mother and my sisters and my cousin's wife and all my dead grandparents—always thought was my inevitable segue from late-night *Seinfeld*s to mornings with *SpongeBob* until very recently. Kind of amazing that a couple with seventy years of combined living hasn't taken a stand on the most natural thing available to us: procreating the species.

Not really. First of all, I'm a guy, which means I'm quite comfortable avoiding all gray matters until I absolutely have to deal with them, an MO that has worked just fine to date. More important, Piper has always made it clear that her dream was never for a man to whoosh her off her feet, shove a rock on her finger, and start making babies. She's the child of divorce, which doesn't make her any different from half the people born in the late 1960s, but it does make her think a little harder before starting a family. I'm not the child of divorce—in fact, there have been few divorces in my family—but it seems obvious to me that the high divorce rate counters the prevailing notion of family unit as Holy Grail. Despite the social pressure to procreate, studies show that people with children are exactly as happy as people without children. Happiness comes in many varieties.

That's Piper's point. I can see it. Maybe kids don't complete us—or at least aren't the only road to a full life. This isn't a political statement. This isn't a "we're too hip to be conventional" sentiment. But it is an opinion that bugs people. We know plenty of couples that have gotten married, had children, and are now strangely defensive about their own choices (though I think that has more to do with the 'burbs than the babies). We have other friends who desperately want children, but haven't found the right partner and somehow resent the idea that we've found each other and yet still might actually pass on parenthood.

Piper believes that we—or at least she—can be plenty happy if it's just us. Time was, you had to have children—more hands on the farm and all that. Those days, and that necessity, are long gone. Our fortress can stand on four legs. It's a logical, yet relatively unspoken idea, especially coming from a woman. In fact, more and more women feel this way—but the reality remains that we're talking about very recent history versus the way things have been for the last forty thousand years or so. Piper has no problem saying: I love kids, but maybe I don't need to have one of my own.

When the rest of the world demands to know why you don't want kids, Piper's response is that people should know why they want them, not why they don't. And if you want them, can you handle it? Are you ready? More specifically, she wants to know, am I ready? Am I, in her words, "emotionally prepared"?

Although my fiancée's official party platform states that she desires no kids, this "emotionally ready" bit is interesting. She'll consider kids, she's said for a few years now, once she's convinced I have a more nuanced grasp of what I'll be getting myself into. But as the years go by, the "emotionally ready" gap closes (if slowly). I'm older and calmer and—yeah—wiser than I used to be. I'm smart enough to know just how much I have to learn. I'm experienced enough to realize that people rise with the celebration or calamity they're confronted with. Nobody is ready for landing the winning lottery ticket or surviving a tsunami, but when it happens, you just deal. Into this sociological stew comes the math: Piper's being 30 percent in favor of something is a long way north of zero. The womb, it appears, is ajar.

In the end, I know I'm just a chicken. I'm as afraid not to have children as I am to have them. I've got a nagging feeling that life could be passing on by, a worry that my experiences are not going to equal total fulfillment. I don't want to be the guy who is so focused on his career and pursuit of what I have traditionally found to be pleasure that the rest of my life happens passively in the background. But when is it ever a good time to mess with a good thing if you've got it? And given that if we do have kids, it will kind of have been my idea, there's no way I'm going to not be majorly involved in raising the things. I'm all for taking over traditionally female responsibilities like play dates and poopy diapers, but shifting my career into neutral, or worse, reverse? I want to be Zen about the future and invoke the spirit of Joseph Campbell who wrote, "We must be willing to get rid of the life we've planned so as to have the life that is waiting for us." Problem is, I'm not sure if I need to abandon my original plan (a family) or the alluring new plan Piper puts forth (an infinite table for two).

There are reams of data available on women's choices, and next to none on men's. In one of the few official looks at male decision-making on having babies, a University of Montana study called "Men's Experience of Making the Decision to Have Their First Child," found that men talked mainly about their fears of what they would lose if they had a child: freedom, independence, and intimacy, for starters. "I was really struck at how little difference there was in how they talked about these potential losses, whether the man was 18 or 40," says study coauthor Dr. Andrew Peterson. So, there you have it: when we think about fatherhood, we don't think about what we'll gain, but what we'll lose. That's one sad statement. One that sounds awfully familiar.

It gets worse. Keep drilling down, and a cost-benefit analysis of children doesn't yield a net gain. Kids are expensive, a constant cause of worry, and—if all goes well—will be completely crushed when you croak. "It's hard to understand why you would ever do this," says a new mom who went through a similar soul-search before getting pregnant. "Then you see your child stand up for the first time, and . . ." Yes, yes, and it's the Greatest Thing Ever.

I'm sure I'm at once too cynical and thinking too hard. But how will I find closure to such a huge choice when, after weighing outside factors—family, history, biology, economics, my pals and peers, Piper's commitment—I still find myself in a very real internal struggle? I keep waiting for the magic moment to happen. But it's not arriving via Ofoto, one of many Web sites I'm alerted to by my rapidly spawning friends eager to share the latest drippingly adorable candids of their kids. It won't come from the many preachings of my older sister (two young boys, one newborn girl, and one hopes a vasectomy to be named later), who enjoys cornering Piper at family gatherings and declaring, "I just want you to know if you have a child out of wedlock, that's okay." It's not arriving on the plane in which I write this, trying to drown out the truly awesome shrieks of the two-year-old a few rows behind me.

Maybe the eureka moment will arrive if Piper decides it's something she wants to do (or at least moves her percentage north of 50 percent). She may joke about that mythical twenty-seven-year-old waiting to procreate with me, but when I toss it back at her, wanting to know what she'll do if I decide I must pursue a daddy destiny, she sighs and admits, "Oh, you'll probably convince me to do it."

But I don't want to if she's not totally on board.

I'd probably love having kids. The question is: can I live without them? I've done a pretty good job of living without them so far. Trouble is, no matter how much the world changes, despite all the technological progress we've made, when it comes to kids, there's still exactly one way to find out.

LAKSHMI CHAUDHRY

Next Stop, Motherland

MY FAVORITE MEMORY OF MY NIECE is of a seriously chubby eight-month-old, humming along as I sing her favorite lullaby, "Karma Chameleon." *Come-a-goo, come-a-goo,* she'd add drowsily at all the wrong moments. I remember her milky, baby-powder smell, the tiny roll of fat around her wrist, the weight of her head on my shoulder, the miniature perfection of her belly button . . .

I love babies.

Yet I wallowed in fear for years before I was finally ready to have one of my own. Just the thought of getting pregnant conjured up nightmarish images of my maternal self: sullen, disengaged, boring, fat. I was terrified of becoming a mother.

In the sixteen years that I'd lived in America, few of the women I'd encountered had been able to prevent motherhood from eclipsing their former lives and selves. Each time I'd tell someone that I wanted to have a baby, the response was always the same: *It will change your life.* Change, I quickly learned, was just a code word for a long list of sacrifices—sex, ambition, fun, time, sleep. Worse, the parents I met embraced this staggering loss with an equanimity that drove me crazy. Their resignation, which suggested that these losses were both a natural and inevitable part of parenthood, left me feeling panicked. Soon, that panic hardened into a fierce, unyielding resistance to having a baby.

My husband, Shaili, grew more anxious with every passing month.

"You don't want to have kids!" Shaili would yell in the midst of one of our interminable arguments about parenthood. "Why don't you just admit it?" I denied it, fiercely, repeatedly, and with great conviction. And yet I was secretly afraid that maybe he knew better. Maybe I really couldn't do it. I knew that there must be women in America who had figured out how to balance the demands of motherhood and their own needs for a fulfilling life, but I hadn't met them. Willingly or otherwise, each mother I knew had sacrificed something—career, a social life, and, in some cases, a happy marriage—at the altar of maternity.

But—to use the ultimate cliché of modern motherhood—I wanted it all. And why shouldn't I? The women I grew up with in India, my sisters-in-law and my friends, were career women who had chosen motherhood without losing themselves. They didn't spend time wrestling with guilt when at work or out with friends, not when they knew that their kids were safe in the care of a close member of the extended family. Not forced to bear the enormous burden of parenting alone, the Indian mothers weren't beleaguered or resentful. Their lives did not shrink into a daily ritual of baby-tending activities, but instead grew richer and more complex, as motherhood added new layers to their identity.

Motherhood in America, however, seemed like a struggle that left women exhausted, isolated, and unhappy. I desperately feared that having a baby here would take me right back to where I'd been nearly ten years before: home alone.

We moved to Santa Clara, California, in 1996, a few months after our wedding, when Shaili took a job at a Silicon Valley technology company. I couldn't wait to escape the academy, having spent most of my adult life in its confines, first as an undergraduate in Massachusetts and then as a Ph.D. student in political science in upstate New

York. With the move, and marriage, I finally felt like an adult. In Silicon Valley, we would get a house, I would find a job. We were going to put roots down in the United States, and finally make our adopted country our real home.

As it turned out, everything about Silicon Valley proved to be a little too real, including the immigration rules prohibiting dependent legal aliens—that would be me—from seeking employment. Forget about a paid job. The laws did not permit me to do an internship or even volunteer. After six years in graduate school, I would have to wait three more years for the green card that would allow me to work. While Shaili was busy at the office, I spent most of my time in our dark, silent apartment. One day bled into the next, blurring in and out of focus like my disintegrating sense of self. There were some classes at the local community college, trips to the public library, and long hours spent cooking elaborate meals that I was too sad to eat. In graduate school, I had been a student, a teacher, a friend, a colleague. Now I felt erased.

The office, at least in Santa Clara, was the source of one's social life, the only form of community. In India, personal connections extend ever outward into a vast network, and are not limited to work. Move to a new city, and someone you know will know someone who is eager to introduce you to their friends and family. No one is too busy to create time for you in their daily lives. There was no web to catch me in the United States.

My years in Santa Clara were a cautionary tale about isolation and my inability to cope with it. I feared that having a baby, even if I worked part time or full-time, would require too much time alone at home. I worried I would struggle with depression and a crippling loss of identity, just as I did when I first moved to California. And I wanted to do anything to avoid that.

In the course of our many conversations about having a baby, Shaili had done his best to talk me out of my fears: we would get help; he would work from home two days a week. But his assurances didn't make me feel better. All of the American mothers I knew, no matter how enlightened and politically aware their partners were, ended up doing most of the work. They were the ones who rushed home from work to relieve the babysitter, planned play dates, scheduled doctors' appointments, and took time off to tend to a sick child. Sure, the nanny or Shaili would help out, but I knew that the job of raising the kid would be primarily my responsibility.

After three years, my green card finally arrived, and Shaili and I moved to San Francisco where I started interning at a magazine. At thirty-one I was older than most of the staff, but I didn't care. I stayed out late at bars with my new colleagues, having heated conversations about politics, and went on reckless shopping sprees with my new girlfriends. I felt alive again. A few weeks after I was hired full-time as a staff writer at a technology Web site, I shimmied around the living room naked, with my first paycheck between my teeth.

Shaili and I agreed to wait until I had established myself as a journalist before we tried to get pregnant. Finally, my depression was over, and we could be blissful newlyweds. Why not treat ourselves to a few years of the carefree coupledom we had been denied?

But the B-word reared its head again two years later. I had just quit my job and was looking for something more suited to my interests. "Next year," I told Shaili when he suggested that we start thinking about having a baby. "You're thirty-four," he reminded me, before launching into an analysis of my declining reproductive prospects. That Sylvia Ann Hewlett chose that moment to publish *Creating a Life: Professional Women and the Quest for Children*—a fear-mongering book filled with examples of women who had lost out on the chance of motherhood because of their career ambitions—just added fuel to his argument. Her appearance on *60 Minutes* sparked a full-blown battle, accompanied by tears, threats of divorce, and slamming doors.

"You should have told me you didn't want children before we got married!" Shaili screamed.

"Maybe I don't want to have children with you!" I spit back.

But that wasn't it. I just didn't know how to be a mother in America. I was raised by all of my relatives. At one point, when I was growing up, there were nine of us living under the same roof—grandparents, parents, two brothers, sister-in-law, niece, and me. My mother, while still the most important person in my life, was not the only person I relied on for care, comfort, food, or play. How could Shaili and I even begin to imagine raising children alone?

Hoping to read my way out of my fears, I turned to books about parenting in search of a blueprint, a formula, anything that would help me learn the secret of happy motherhood. But what I read made me even more scared. *A Life's Work* by Rachel Cusk described motherhood as "an unpeopled continuum" shaped by "hours of darkness." An anthology, *The Bitch in the House,* was filled with stories by exhausted women wrestling with rage. The extent of their misery took me by surprise. I tried to remember the mothers I knew from back home. Had I coasted through life in India, merrily oblivious to the legions of overworked, unhappy mothers around me?

No. I didn't remember motherhood as drudgery in India because it wasn't—at least not for educated, middle-class women. The extended-family network made child-rearing a shared responsibility. My sister-in-law, Dharini, a manager at a clothing company, happily went back to work six months after giving birth to my niece twenty years ago—long before a full-blown career became a legitimate goal for middle-class Indian women. Every morning, she handed off my niece to the maid, who would then tend to Shivani and her own two-year-old until noon. My mother and grandmother would hold down the fort until I got back from school in the afternoon. Then the baby was in my charge until her parents returned in the evening.

This isn't to say that Indian mothers lead contented, idyllic lives of comfort. The close familial bonds present their own set of problems. Sharing the burden requires ceding control, be it over what to name your baby or when she goes to sleep. The task of managing the various personalities of family members in the interests of domestic harmony is sometimes more daunting than tending to the needs of a single child. This is a culture where the need for "personal space" is viewed as a puzzling eccentricity.

But despite its downside, the hypersocial nature of Indian culture makes it perfect for the business of child-rearing. Indian homes are not private havens but bustling semipublic spaces, filled with relatives, friends, and neighbors. Every May, when I was growing up, my aunts would descend on our home for weeks, with their children in tow. My mother set up camp in the kitchen, gossiping and cooking with her sisters-in-law while we children played out on the street. My cousins and I spent nights on the living-room floor, where my mother laid out rows of mattresses. All day long, we smuggled food out of the kitchen and saved it for the "midnight feast" we had after our parents went to bed. We'd giggle, wrestle, play until we fell asleep in the early morning hours, only to start all over again the next day. These are some of the happiest memories of my childhood.

In India, children are everywhere. There is no escaping the crying babies in the movie theater, the toddlers stumbling around the French bistro, the clutch of high-decibel kids playing on your street in the evening. And everyone—surly teenagers, hip singles, and crabby senior citizens included—is happy to have them around. Take your baby to a party in India—or to an all-Indian get-together in Silicon Valley, for that matter—and the chances are you won't see her until the end of the evening, as she's passed from one doting stranger to another.

Raising children in America seemed so daunting to me because of all the barriers between the adult world and the children world. Sitting at my favorite restaurant in San Francisco one evening, I noticed that there wasn't a toddler in sight. At a fund-raiser, I watched people admiring the baby of a colleague from a safe distance and then moving quickly along. At a party of a friend, parents hustled their two sons out the door as soon as they started chasing each other up and down the hallways. "We should go before we outstay our welcome," the wife said before leaving. Americans are fiercely devoted to the idea of children, but they have little patience for the messy, real-life, flesh-and-blood version—unless the child happens to be their own.

After eight years of marriage—after too many fights, and too much reading and observing, I told Shaili the truth. "I can't have kids," I finally said. "Not like this."

As it turned out, he had been moving toward the same conclusion. We were ready for parenthood to change our lives, but we didn't want to sacrifice our lives to it. Just as we were both coming to this realization, we went back to India for a long-overdue vacation.

In New Delhi, I hung out with Shaili's sister-in-law, Janaki, a high-powered executive at a multinational company, who has three children under the age of nine. It was morning, and we were in the living room, chatting aimlessly, while Janaki's husband bathed the kids. She was describing the power dynamics in her office, when her three daughters came flying into the room, insisting on trying on the new clothes I had brought for them from America. Janaki spent a few minutes trying to persuade them to wait until later and then finally called out to her husband: "Depi, can you handle this before they drive me crazy!"

There wasn't a trace of guilt or anxiety in Janaki's voice then or later in the week, when she called to say that she wouldn't be home until very late from work. Why should she feel guilty? She knew that her kids didn't rely on her to provide for their every need, she knew that there were plenty of relatives around to help them brush their teeth, tuck them into bed, step in and take up the slack.

During the course of the vacation, I was part of this tag team. I loved dragging the little monkeys around, the twins hanging off each arm, while the eldest, Aishu, ran ahead. We went shopping, painted each other's toenails, watched *Beauty and the Beast* three times. I'd forgotten how much fun kids are—even with the loud tearful tantrums, heedless vandalism, and bitter squabbling. The girls were happy, too: They had grandparents who spoiled them, a new playmate in their aunt, a big, rowdy gang of friends who ran in and out of their apartment, and, of course, their parents.

So, after spending years waiting for my green card, I am moving back home. Shaili recently asked to be transferred to his company's office in Bangalore for at least two years after we have a baby. We're going to start trying to get pregnant soon.

For the first time since I began contemplating motherhood, I feel relaxed and confident. I won't be alone when I have my baby. Of course, I now have a whole new set of fears. I worry that the communal culture in India, which makes it such a wonderful place to raise a family, might feel stifling. I worry about finding a job there that I will like. And I worry that when I return to San Francisco in a few years, the friendships and the career I worked so hard to build will be gone.

"It will change your life," the American parents say. And they're right. Having a baby will do exactly that, but not in the way I expected. For me, motherhood will bring with it a brand-new life, tens of thousands of miles away, in the homeland I left years ago. All I can do is hope that in changing everything, I will be able to hold on to the parts of myself I want to stay the same.

JOE LOYA

Redemption

WHEN I WAS TWENTY-SEVEN YEARS OLD, I told my cellmate, Lalo, that I was never going to have kids. I was at Lompoc Federal Penitentiary, a maximum-security prison in California, beginning a seven-year sentence for bank robbery. We were locked in our cell at night, a cell the size of a parking space. Lalo was on the top bunk, talking about how much he loved his young boys, telling me I should have kids, too.

"Are you crazy?" I said to Lalo. "I'm a thief, a criminal." I could barely wrap my head around all the time I had to serve and was feeling lonely, pathetic, and scared. But I was clear about one thing: I would be a fucked-up father. I was reckless and violent, and had been in and out of jail for years. As far as I was concerned, Lalo was a total failure as a father. He had two children while committing crimes, and had abandoned them to do this fifteen-year hit in prison. "You think leaving kids at home while you do time is a good thing?" I said. "You've gotta be nuts."

The truth is, I'd known since ninth grade that I would never have kids. I grew up in East Los Angeles, and the way my childhood played out—with death and brutality and sexual molestation all around me—I just assumed that any kid of my own would go through the same hell. I had already lived through it once and didn't want to go through it again.

My parents had me when they were both sixteen. When I was nine, my mom died of kidney disease. I was shocked by her abrupt disappearance, left feeling empty. She had abandoned me. Three years later, a twenty-two-year-old female neighbor seduced me. We snuck around and had sex for several years. In the beginning, it was fun. But over time, I began to realize that she had taken advantage of my desperate need for attention after my mom died.

Around the same time, my dad turned brutal. He sometimes beat the shit out of me and my younger brother. I don't mean he slapped me or hurt my self-esteem with scalding remarks. I mean he was a sadist. He once tied a belt around my brother's neck, lifted him off the ground, and spit in his face. Other times, he made us pose in the push-up position and hit us with a baseball bat when we tired and fell to the floor. He punched us in the mouths and kicked us in the ribs when we were down.

But things changed when I turned sixteen. During a particularly vicious beating, in which my father fractured my rib and elbow, I stabbed him in the neck with a knife. He survived and never hit me again.

In high school, some of my friends talked about having kids. But not me. I knew that my rage, like my father's, could easily spill onto my own son. And truthfully, a part of me feared that my son would be willful like I'd been, that he could one day grow up and stab me in the neck in order to gain his own freedom.

After high school, I worked in restaurants and tried junior college. But I had no concentration and dropped out of both work and school. I couldn't see myself working through college, training at a good firm, slowly working my way up the ladder to a high-paying job. I didn't have the patience for all that. So, at twenty-two, I began a life of petty crime, bouncing checks, stealing, and conning my friends. Then I graduated to stealing cars and strong-arm robbery. I went on a fourteen-month bank-robbery spree, hitting over twenty-four banks in southern California.

I was full of rage during my crime days. One night, I was watching a Lakers game on TV with my friends. When the Pistons led by twelve points in the fourth quarter, I got up and threw the TV against the wall, smashing it to pieces. As my friends sat there petrified, I walked calmly to my bedroom and brought out another television.

I was finally caught when a teller handed me a stack of bills with a transmitter in it. The police tracked me down seven miles from the bank in Los Angeles. I was sent to federal prison, where I continued to commit crimes like gambling, making knives, brewing prison wine, and smuggling drugs.

Two years into my prison sentence, Lalo was stabbed to death in his cell as he slept. We hadn't been cellmates for months, but I was swept up in the homicide investigation with five other Mexican men. We were locked in solitary confinement, under investigation, for almost two years. I went partially crazy. I imagined the cell walls were moving. I heard voices. One night, I hallucinated that I was being visited by a short, bald boy. I awoke from that vision terrified that I couldn't tell the difference between reality and fantasy.

When the administration finally figured out who the real murderer was (none of us), we were placed back into the general prison population.

But that long period in solitary broke my criminal spirit, and I began to imagine a life outside of prison. I was so tired of it. I hated having no say in the food I ate, the clothes I wore. I hated having my movements restricted. I hated the everyday humiliations, like receiving only one roll of toilet paper a week. God forbid that any one of us should get diarrhea and go through our single roll in three days.

But more than the tediousness of prison, I wanted to change. I'd been raised in a fanatically Christian home, and so, stories of redemption were not new to me. I had squandered my life and hit rock bottom. At my absolute lowest, I felt humbled, like the prodigal son, and began to think about reform. I was willing to try and understand my rage and insecurities. I knew I would never rob banks again.

Toward the end of my last year in prison, I found it hard to imagine any woman would want to be with me, much less have children with me. In one of my final letters to a friend, I wrote that I expected to remain single for the rest of my life, that I felt disqualified from ever being in a healthy relationship.

But when I got out of jail, I learned that I could attract women, even women who were willing to start a family with me. I had girlfriends who said they felt safe with me. They told me I was the marrying kind, because I was patient, a good listener, and I asked a lot of questions that made them feel seen and heard. For the first time in my life, I was told that I was "present" and "mindful" in a Buddhist sort of way. (Some Buddhist pamphlets had indeed found their way into my cell, and they did influence my thinking about myself and the world.)

But I wasn't sure if my change took. I still brooded a lot.

Then I started dating a woman who was bipolar. Her sister was schizophrenic. A crazy uncle had killed himself. I began to wonder about the moral implication of passing mental illness on to a child. Wouldn't that be like having AIDS, knowing full well your child would be born with the disease? I pondered the issue for a while but let it go after I broke up with the woman.

Near my ninth month out of prison, I began having fantasies of killing myself and went to see a psychiatrist. He told me that I was chronically depressed, and prescribed drugs that made me chatty and personable. No more brooding Joe.

At the time, I was living in Los Angeles, and one night I had a birthday party. My friend Ofelia showed up with a woman named Diane, who was visiting from the Bay Area. As the night progressed, I liked what I noticed about Diane. She listened well and asked thoughtful questions with a kind voice. The party lasted until morning, and afterward we drove to eat breakfast at L.A.'s famous all-night diner, the Pantry. When we pulled up, Diane displayed her talent for parallel parking. I liked her immensely.

I was now at the top of my game. I was writing a memoir, working as the L.A. correspondent for Pacific News Service, having my op-eds published in the *Los Angeles Times*, and making friends. It felt good to be living an honest life. One year out of prison and seemingly out of the dark woods, I felt optimistic about my life.

Three months after I met Diane, I moved to Oakland to be near her, and we started dating. I loved Diane in so many ways. I loved her work—she was a program officer at a foundation that funded non-profit community health clinics. I loved her humility. I even loved the way she stood at the sink when she brushed her teeth—erect, like the ballerina she once was.

With Diane, everything was new and possible. Being with her made me feel for the first time that I could be a good husband and father. We had conversations about what kind of parents we wanted to be, and the conversations didn't scare me. The idea of having children with Diane felt natural and right.

After dating for eighteen months, we got married. We bought a house in East Oakland, got a dog named Olive, and purchased a VW Passat station wagon. We were preparing a home and life for children. After four years of marriage, we decided to start trying, timing the baby's birth with the completion of my memoir.

But things didn't go as planned. I wasn't close to finishing my memoir and constantly felt pressured by the encroaching deadline. Two weeks after the book was due—I still hadn't turned it in—one of my dearest friends died in a car crash. The grief of his loss paralyzed me and brought up a lot of my ancient grief for my mother.

Diane and I were also experiencing a serious financial crunch. I wasn't bringing in any money with my writing and was feeling terribly inadequate. My stress was compounded by an onslaught of ferocious suicide fantasies. In one, I saw myself walking into my kitchen, choosing a large knife, and cutting myself open from sternum to crotch, zigzagging the wound all the way down.

These hallucinations were so vivid and jarring that I told Diane we had to stop trying to have kids. Even if I didn't kill myself, I didn't want to pass on this mental illness to our child. I didn't want our children to be tortured by the deep lows I was now suffering. I didn't want them to fantasize about carving their chest open with kitchen knives.

Diane was heartbroken. She felt that she had been cheated. She had married a man who said that he was willing to have children, and now I was telling her that I wouldn't. I suggested that she leave me and find a man who wasn't as fucked up as I was.

I wished I'd never met Diane. I despised the soap-opera quality of my suicidal tendencies. I felt like an attention-starved actress. I hated that my life had turned into an after-school special. Most of all, it was horrible to know that I was causing Diane so much grief.

Once, at our lowest point, Diane cried out, "Okay, I won't have children. But you are going to have to pay for all the therapy I'll need to get over it." Tears flowed, and our house was full of dread and silence.

Diane was pissed off, and I understood that. But a big part of me felt misunderstood. She was angry because she couldn't have a child and saw me as being selfish in my refusal to have one. But my thinking wasn't "Let's not have kids so that we can live unencumbered lives with lots of travel and fine dining." I was twisted up in a psychological knot and feeling strangled by self-hatred. My thinking was more like "Let's not have children, Diane, because the stress of it might send me over the edge and then I'll kill myself and leave you a widow with a newborn child." So, Diane's pissiness irked me and made me suspect that she didn't really understand the gravity of my condition. At times, I felt like she wanted to say, "Give me the child I've always felt entitled to, your mental health be damned."

During those hard, strained months, my suicide fantasies hit with more frequency. One day, while driving on the interstate from Los Angeles to Oakland, I imagined my car crashing into a pole, my head bashed in. I unbuckled my seat belt, sped up to 110 mph, and started sizing up the poles at the side of the road. Finally, I scared myself enough to pull off the highway and park at a gas station, where I slept for two hours. When I got home, I told Diane that I was falling apart. The next day she helped admit me to a psychiatric hospital, where I was diagnosed with a bipolar disorder.

The hospital was a strange place. Doctors gave me a combination of drugs that made me sleep all day. My mind had royally screwed me, but I was still the sanest person in my ward. One morning at three a.m., my schizophrenic roommate woke me up out of deep sleep, cursing and carrying on a very loud argument with the devil.

While I was in the hospital, I thought a lot about children. I couldn't foresee a time when I would want them. In fact, I felt the choice had been taken from me. Why would someone with a debilitating mental disorder ever subject kids to the same condition? I believed that not having children was the moral thing to do.

After eight days, I was released. I felt fatigued and tender from the months of extreme confusion, but I didn't want to lose my wife. She was the only stable thing in my life. I told Diane that our marriage had been solid before my mental collapse and deserved the chance to be strong again. We were exhausted but didn't want to separate. We agreed that we would take a year to try and salvage our marriage.

While my bipolar problem had an easy solution—pills—our relationship was another matter. Our home was still a painful place, because Diane wasn't sure I was completely well. She was afraid that I would crack at any moment. We could wish for a best-case scenario. But Diane knew that so much of what she wanted to accomplish was out of her hands and in the hands of something more precarious—my mental health. Sometimes I would hear her crying in the next room.

It was a rocky time for me, too. It was hard to face the damage I had done to our marriage. Before I went to prison, I had emotionally terrorized every girlfriend I ever had. When I thought about changing in prison, I swore to myself that I was never going to make people organize their lives around my moods again. Now I felt I had resorted to old habits, Bad Joe up to his old tricks. I felt that I was a big fat failure at reform, a hostage to stupid hope, and quite possibly a fraud. In my worst moments, I couldn't shake the feeling that suicide seemed the suitable end for such a pathetically bungled life.

But over the course of five months, I began to regain my form. The pills were allowing me to feel like my good-natured self again: playful, peaceful, and secure in the fact that my mind wouldn't betray me. I felt that I had turned a corner.

And Diane did, too. She could see how badly I wanted to repair our shattered trust. I was being extra loving, lingering longer in a hug, kissing her more often, and spending more time in her arms before we went to sleep. My bipolar condition was in front of us, and as long as I stayed on top of it, with weekly visits to my psychiatrist, who monitored my drugs, we felt like we wouldn't be blindsided by my mental illness again. Doom and gloom were replaced with tentative optimism.

More than anything, I felt far away from suicide. I didn't feel it in my body anymore. My fantasies now seemed like they happened to another person, much the way my robberies felt like someone else committed them. I was no longer full of rage but in a safer, tamer place.

A year and two months after I checked into the psychiatric hospital, Diane and I started trying to have a child. Our marriage took a hit and we hung tough, we survived. And this made us want to pass on more love. That's how I view our future child: a person born from a union of great love and survival. A defeat of the ugliness I have endured in my life.

Today, I'm forty-three. My home life is a far cry from my traumatic childhood and stressed-out years in prison. Diane is an amazing partner, willing to talk through our problems. We are considerate with each other and I feel largely at peace. I am in the most tranquil space of my life.

It's been a long process, but I've overcome my fears that I ever would beat my kids, go to prison, abandon them, or pass on a mental disorder. I'm optimistic that I'll be a good father—knowing, alert, and good-humored—and if nothing else, I'll have some really great bedtime stories to tell my kids one day.

PART THREE

Taking the Leap

AMY BENFER

Parenting on a Dare

MY DAUGHTER WAS FOUR DAYS OLD on the day I decided to be her parent. My father was in the driver's seat. I was a sixteen-year-old, leaking milk in the backseat of my parents' station wagon when I made the announcement: "Let's go get her."

"We're going to get the baby," my father said, and drove over a wall of traffic cones to cross over into the turn lane. And that was that. Me, my parents, my younger brother, all of us went on this reckless mission to pick up a newborn baby we'd left with my mother's single friend until we figured out what the hell to do with her. I don't know how my father got there without killing us all. We were all crying. We knew it was a stupid idea. We knew we could be seriously messing up at least two lives. We knew that there was a perfectly appropriate adoptive couple, of the right age and financial situation, with a mortgage, two cars, and a nursery waiting for that baby. But somehow we all rushed into the friend's condo, claimed our baby, and took her home, where she went to sleep in a tiny crib that had housed my baby dolls not that many years before.

We all made promises to that child. My brother, who was ten at the time, promised to donate all of his allowance for the next eight years, and a new pair of Bugle Boy jeans if we'd keep her. (We still haven't collected on that part of the deal, but I remind him every few years that coming through with at least the Bugle Boys would be the right thing to do.) My parents promised to get me somehow to adulthood, by providing us both with a place to live and health insurance, giving me free child-care (from my stay-at-home mother) through high school, and sending me on to college, as they had already planned to do.

My promise was the most complicated. On the morning my daughter was born, the doctor came in to talk to me. He knew my age. He knew there was an adoptive family waiting to take her (the bouquet they sent was on my nightstand). He knew I had not yet made the decision.

He told me that he was concerned with taking care of his patients' minds as well as their bodies. He said that, of course, the best decision for my daughter would be to place her for adoption. But, he said, perhaps I was not strong enough to make that choice.

My response? *Fuck you.*

My doctor meant well, as did everyone else who had said more or less the same thing throughout my pregnancy. And I was the kind of girl that everyone believed would recognize that parenting my daughter as a teenager was not in the best interest of myself or my child.

But I took it as a dare. I had a sixteen-year-old's immortality complex. Up until then, nothing had been hard. I had had a safe, middle-class childhood in which the only dangers were those I got myself into—by, say, taking the family car in the middle of the night, or having sex.

I hated my doctor for implying, or so I thought, that keeping my child would be a sentimental decision made out of emotional weakness. To me, this was my chance to prove just how tough I was. Like most ambitious teenagers, I still believed at that time that I could choose between, you know, being president, winning the Pulitzer Prize for fiction (maybe poetry, maybe both), and maybe being a movie star or something. I was still going to do all that, and I was going to do it with a child. It was my chance to be extraordinary in the most literal sense, by breaking out of the ordinary college, career, dating, marriage, children trajectory that was expected of girls like me.

So my promise, to myself and my daughter, was that we were going to prove everyone wrong. I was going to raise a child who was every bit as smart, capable, and badass as I thought I was. And I was going to do everything I would have done anyway, and do it just as well as, if not better than, I would have on my own. It was the Enjoli commercial, with a baby.

My mother and I, both of us big talkers, would have long, ponderous discussions on why, exactly, teenage mothers often did badly. We didn't know many of them, and none intimately. We came up with theories, some more crackpot than others. We took our cues from novels, from people around us, from pop psychology. Of course, there was the money thing and the education thing. But we wanted to understand the psychology of teen parenting. One of us—I'm not sure which one—came up with the idea that teenage mothers were often emotionally stunted at the age when they first became pregnant, because, as we decided, they hadn't "gone through all their developmental stages."

It was an arrogant pronouncement, one that certainly revealed that neither one of us had much experience with any situations that fell outside the range of typical family life, the kind of life we had just agreed that I was going to have.

I took it to mean that, at sixteen, I should act like a sixteen-year-old; at eighteen, like an eighteen-year-old; at twenty-five, like a twenty-five-year-old, and so on. The danger, as we saw it, was that if I gave up too much of my own identity to being a mother at a young age, I would resent my child and that would be a bad thing.

In other words, being a good mother, for me, was entirely dependent on how good I was at taking care of myself. I not only gave myself license to be selfish, but following my selfish instincts also became a moral imperative.

For the first two years of my daughter's life, my parents insulated me from the usual consequences of teenage motherhood. There was no question that I was my daughter's parent, but I certainly was not a single parent. My mother did the child care from eight a.m. to three p.m.; my father earned the income. My job was to do well in school and to enjoy my baby. And since my job was also to be a "normal" high school girl, I went out with friends and to punk rock shows on the weekends, after my daughter was asleep, and to poetry club at the coffeehouse every two weeks. I was too busy to date much until the end of my senior year, but if I'd wanted to, I could have done that, too.

Because it was so easy, I got bolder. The first major conflict between my parents' generosity and my own ambitions came around the time I started to apply to college. My parents, who had both gone to state schools, had once been perfectly happy to send me to whatever private school I wanted to attend. They still agreed that I should go to college, but they wanted to keep me close to home, where they could help out. I saw no reason why I should scale back my ambitions just because I had a child. I applied to all the same schools I would have anyway, and finally chose a very expensive one three thousand miles away from home.

My parents were horrified, but in the end, they didn't stop me. I don't think anyone could have. When they said it was too expensive, I asked for their tax returns and suggested ways they could meet the expected parental contribution. When they worried about how we would live, I called the dean of my school and came back with child care, a two-bedroom apartment, and a meal plan for us both. Two years later, when they told me they could not afford to pay for that school, I took a year off and went back as an independent student. My daughter and I moved to Connecticut when I was eighteen, and, except for brief visits back, I have never gone home again.

Every few years, I try to write an essay with the working title "Without You." It's supposed to be a piece that follows the imaginary person in my head, the version of myself who has lived a parallel life without taking that dare and deciding to raise a child at sixteen. I've never been able to write that story. Part of it must be that having a child shapes every part of a person's identity, so that it becomes impossible to imagine a self who has not been formed by taking care of that child. But the other reason I've never been able to write that story is that having a child didn't change me *enough*.

This year, I will have been a mother for exactly half my life. I'll turn thirty in less than a month; my daughter will turn fourteen this summer. If you had asked me at fifteen what I saw myself doing at thirty, I would probably have said that I would be a writer living in New York. And, at thirty, I live in Brooklyn and have made my living as a writer for nearly seven years. At eighteen, I was an English major at a college I loved. At twenty-three, I moved to San Francisco to be near my boyfriend, a brilliant novelist whom I loved. At twenty-five, I had my dream job as an editor at *Salon*, which, yes, I also love. And at twenty-nine, I moved to New York, where I am still doing work that I love. I'm not going to pretend that I couldn't write another version of my life about all the ways I've failed to get and do what I want. But it's difficult to believe that the bare outline of my life for the past fifteen years would look any different if I'd had the freedom to construct it without taking my daughter into consideration.

I kept the first part of my promise to myself: Being a parent didn't stop me from doing what I wanted to do. The second part, is, of course, my promise to my daughter. Fifteen years after she was conceived, do I think I did right by her?

The worst part of being the parent of older children is that you see the effects of everything you have done. You know exactly how you have damaged your child, in the same way that you know exactly how your own parents have damaged you. It's a cost/benefit analysis. I don't always know what she thinks of me, but I know what kind of parent I am. I still have a nearly subhuman immunity to risk. After you make the decision to raise a child at sixteen, few things compare. Nothing seems crazy to me. I am a warm parent. I love to talk to my daughter. But I'm also undisciplined. I'm messy. I'm selfish. We've never had enough money, enough space, enough time. When I fail—to clean the apartment, to meet a deadline, to find a job, to keep a lover—I have the same tendency to think *fuck you.* You try to raise a child on your own at sixteen (at twenty-two, twenty-five, twenty-nine). It's an excuse. People let me use it more often than they should.

I am in the strange situation of having known the couple who wanted to raise my child if I had not chosen to do so. I selected them, I sat in their living room, I toured the house and talked with them about their philosophy of child-rearing. We haven't spoken in fifteen years, but I've heard enough to have an idea of where they are now. I know that in her parallel life, my child would have grown up the daughter of a social worker and a real estate agent in the Northwest. She would live in a four-bedroom house; she would have a summer cottage in the mountains. She would have been allowed to have a dog, and it's likely she'd have a brother and a sister. She wouldn't have seen as much of the world. She may have grown up to be more like the girl I was at her age, a suburban teenager longing for adventure and danger and a more exciting life. She would have been more stable.

I don't know that girl. She isn't my daughter.

The girl who is my daughter has been told her whole life that we almost didn't do it, that we almost lost our nerve. She knows as well as we do that was the sensible thing to do, and we haven't tried to hide it. As she's gotten older, she and I sometimes talk about what her life could have been like. She can't really imagine it, of course, any more than I can. I have an odd little line that I trot out sometimes. When she says she was an "accident," I tell her that she should feel that she is all the more a wanted child. We didn't want a child, I say, we wanted *you*. She wasn't convenient, she wasn't planned, she profoundly changed all of our lives. And we did it anyway.

I say "we" because I am thinking of the four of us in that station wagon, running over traffic cones on our way to do something that everyone else knew was a terrible idea. And you know what? We were exactly, exactly right.

DANI SHAPIRO

Not a Pretty Story

ONE DAY, WHEN I WAS NEARLY THIRTY, I became aware of the details of my own conception. My mother had come to visit me in graduate school, and I had just introduced her to my friend Rachel.

"And where are you from, Rachel?" my mother politely asked.

"Philadelphia," Rachel responded.

"Oh, really," said my mother. "My daughter was conceived in Philadelphia."

This bit of salient information slipped sideways into my life, the way so many bits of information did in my family: bombshells delivered breezily, casually, seemingly unimportant enough to share with a perfect stranger. *You were conceived in Philadelphia. Please pass the salt.*

"I was?" I asked. I had never really wondered about where I was conceived. That would have meant imagining my parents having sex. But now I was curious.

My mother nodded, as Rachel looked on.

"Where?" I asked. "How can you even be sure?"

"Oh, you don't want to know," said my mother. "It's not a pretty story."

It took a couple of hours to pry it out of my mother, once we were alone. As I drove her back to New York City late that night, I finally convinced her that she couldn't just tell me that the story of my conception wasn't pretty. She couldn't leave it at that.

"You were conceived by artificial insemination," she said with a sigh. I focused on the road, both hands on the steering wheel.

"What? You mean, like, in a test tube?"

"Dixie cup, actually."

"They did that, then?"

"In Philadelphia," my mother said. "I used to call your father on the floor of the New York Stock Exchange—"

Inwardly, I rolled my eyes. My mother loved that phrase, "on the floor of the New York Stock Exchange." Even though it was just the two of us in the car, even though my father, who had been dead for several years, had worked on the floor of the New York Stock Exchange for my whole life, she said it grandly, as if to impress.

" . . . and I used to say, 'Paul, it's time. Get on the next train down here.' And your father would drop everything and make a mad dash for the train."

I could tell that my mother was now enjoying telling the story. She had forgotten that it was a story she had kept from me for nearly thirty years, and was now milking it for the drama.

"Why?" I interrupted. I knew my mother had had miscarriages, more than one, but the problem hadn't been fertility.

"Slow sperm," she replied.

"So he—I mean, Dad—"

"They gave him a magazine and sent him into the bathroom with a Dixie cup," my mother said. "This doctor in Philadelphia—this was 1961—he was the first in the country. I did my research." She paused. "I told you it wasn't a pretty story."

I dropped my mother off at her building on Riverside Drive and drove for a while around Manhattan. It was late at night; there weren't a lot of cars on the streets. I could still feel the cool ledge of my mother's cheekbone, the almost bloodless feel of her skin touching mine as we kissed good-bye. I had always wondered if she was really my mother—not just in that hateful-childhood-fantasy way, but as a real, ongoing question. There were no photographs of her obviously pregnant with me, but that could be explained by the bed rest that she was confined to for most of her pregnancy. I knew that I had been born by cesarean section, and I had seen her scar, but still, I had never been entirely convinced. She and I were nothing alike. She was tall, dark, angular, dramatic. I was small-boned, soft-featured, fair. But our differences, as I saw them, ran far deeper than our surfaces. On a *soul* level, it seemed impossible to me that this woman was my mother. All my life, I had had trouble looking her in the eye. Her presence brought me no comfort. In my longing for some sort of pillowy, maternal warmth, I came up emptier than empty. For years, I had collected other, older women. Mother figures who bestowed hugs, asked questions about my life, gave me compliments that weren't wrapped in barbed wire.

As I drove the empty streets, I tried to adjust to this new information. In a way, it made perfect sense. I was conceived via artificial insemination, then my mother spent nine months flat on her back, then I was cut and lifted out of her, my first touch in this world that of a surgical glove. I was raised as a delicate hothouse flower—or, perhaps another way of thinking of it, I was a scientific experiment. Observed carefully, kept apart, under wraps. Even though I was a perfectly healthy child, I was scrutinized. Poked, prodded, put under the magnifying glass. Every fever was a life-threatening event. Every lump or bump was ominous—a sign, perhaps, that I wasn't meant to be, after all. Was it the Philadelphia doctor's office, the Dixie cup, the bed rest—*we wanted you so much,* my mother used to tell me—or a force darker and more complicated, that made my mother shiver over me so? She wanted me, yes. But she hardly knew what to do with me once she had me. I grew up feeling like there was something, the thinnest, most transparent membrane, but a membrane nonetheless, separating me from the rest of the world.

I never thought I could have children of my own. Though there was, in fact, absolutely nothing wrong with my reproductive system, childbirth seemed somehow impossibly out of reach. Motherhood was the province of other women—stronger women, more substantial women. My breasts weren't meant for breast-feeding. My hips weren't made for pushing a baby out. I could not imagine myself carrying a child to term, giving birth. I had an eerie certainty that becoming a mother would kill me. In my twenties and early thirties, I didn't feel I had a *right* to be a mother. That whole part of life—the mother-daughter part—had eluded me as a daughter, and it seemed only natural that it would also elude me as a mother. But really, that was fine with me. There was more to life than motherhood. Or daughterhood.

By the time I reached the age when many of my friends had started their own families (this age being a relatively late one, given that I lived in New York City), the idea that childbearing was lethal became tangled up with the idea that I didn't desire children. I was thirty-three, thirty-four, thirty-five. If my biological clock was ticking, I didn't hear it. I watched my good girlfriends as their bellies grew; I threw them baby showers and visited them at home with their children, and listened carefully inside myself, as I left, for any hint of envy or regret—and found none.

But then, something happened to me, something I had wanted very much to happen, but that amazed me nonetheless. I met a man—*the* man—I knew I would spend the rest of my life with. I was thirty-five when we married, and still, no tick of the clock. We discussed children, but it was all very abstract. He seemed okay with the idea of not having any. I figured we would probably be one of those interesting, literary childless couples who had disposable income and went on adventures, living the life of the mind. The life of the mind was, of course, where I felt most comfortable. None of that messy, bloody, corporeal stuff. I collected happy (or at least literary) childless couples in my mind, shuffling through them like playing cards. Jean-Paul Sartre and Simone de Beauvoir! Lillian Hellman and Dashiel Hammett!

All well and good. But then, one early spring evening, we visited friends in our Upper West Side neighborhood for a Chinese takeout dinner. These were friends for whom I had thrown a baby shower three years earlier, when their daughter was born. I hadn't paid much attention to their daughter as she morphed from an infant into a sunny, precocious child—a child, who, as it happened, adored my husband. "Would you read me a bedtime story?" she asked. As she led him to her room, they were framed by an archway, and here is the moment that did me in: she reached up a tiny hand to grab my husband's hand, and then—out of nowhere—my eyes were flooded with tears, and I turned away, embarrassed by the suddenness and intensity of my desire.

He needs to have children, I thought to myself. It was a thought with neon lights around it, so shockingly clear that it cut through layer upon layer of doubt. Was it, in fact, my own desire that I was feeling? Perhaps—but I couldn't have allowed myself that. Once I had the thought, it continued to play around the edges of my mind, and no matter how I tried to tuck it away, it always kept coming back and announcing itself. *A baby.* No. *A baby.* No. It was the rhythm of my inner life—almost, but not exactly, like the ticking of a clock.

Which explains the summer afternoon after we'd been married just over a year. My husband and I were making love, and I knew, suddenly and with absolute clarity, that we were making a baby, right then and there. It was the first time we had ever had unprotected sex. Hubris, is it not, at the age of thirty-six, to be so sure? To know so absolutely? And yet, I did know. I did nothing to stop it. Quite the contrary. I forged ahead, blindly and out of sheer love for my husband, figuring that I might not have any idea how to be a mother, but no matter what, he'd be a great father. As we lay there, the late afternoon light filtering in through the rickety blinds of our beach cottage, I silently cheered on the . . . what? Sperm? Zygote? Life forming inside of me? *Come on,* I encouraged it. *It's okay. Come on.*

Throughout my entire healthy pregnancy, my mother kept looking at me strangely, as if I was some kind of alien creature best observed from a distance. There was no patting of my belly, or listening to the heartbeat, or shared mother-daughter information. Never once did I ask her a question. *Did you ever . . . ?* I could see that my morning sickness in the first trimester puzzled her, as did my exhaustion later on. Pregnancy seemed, for my mother, like a food she had never tasted, or a country she had seen pictures of but never visited. When I try to recall her face during those months, I see her head slightly tilted to the side, a strained little smile—but mostly what I remember are her eyes: they were wild, at sea. She looked completely and incontrovertibly lost, befuddled, as if I had broken a pact I hadn't even known existed.

I avoided my mother as much as I could during my pregnancy. I was riding a peaceful, worry-free hormonal wave when I wasn't around her, and she was the one person—the only person in the world—who could disrupt it. Here's how the thinking went, though these words didn't exactly form inside my head—they simply knotted up into a ball of panic: she's not really my mother—or maybe she is, but I was brought into this world by every possible scientific intervention. I shouldn't be here. And I (who shouldn't be here) am bringing another life into this world who shouldn't be here. I didn't have a mother, and therefore I shouldn't be a mother. I have no right.

My son was born by emergency cesarean section after thirty hours of labor, and he was, if I may say, magnificent. My C-section didn't surprise me. At least some repetition felt inevitable. A short while after my husband, newborn son, and I were settled into a recovery room, I asked my husband to call my mother. It was about ten o'clock on a beautiful spring morning. We were in the same hospital where I had been born almost exactly thirty-seven years earlier. She lived a ten-minute cab ride away.

Three hours later, my mother arrived at the hospital. She was wearing her best suit—one she might have worn to a charity luncheon or a matinee—and, even through my Percodan haze, I realized what had taken her so long: she'd gone to have her hair done. I have a photograph of the moment she bent over and peered into the bassinet where my baby lay sleeping. That helpless, lost look is on her face, and as she bends over, her arms are crossed, as if she's shivering, holding herself.

"He's beautiful," she cooed, but her tone of voice was ever so slightly off. "He's perfect," she said. But it sounded to my ears like a lie.

For the first time in my life, I let my guard down. From the moment I discovered that I was pregnant until my son was six months old, I lived in a world unfamiliar to me: a world in which catastrophic things didn't happen. I went from being sure that pregnancy would kill me to a blissed-out certainty that everything was—and would continue to be—all right. I was not worried about all those bad pregnancy conditions like gestational diabetes, preeclampsia. Even during my emergency C-section, I was largely unafraid. I continued to float along even after Jacob's birth. I did not check his breathing every five minutes. I did not freak out when he coughed, or spit up, or cried. I was not going to treat him like a fragile creature. I was not, under any circumstances, going to be anything like my mother.

And here was another first for me: I stopped speaking to my mother for weeks at a time. It was as if my own new motherhood had created a cocoon around me, offering me protection from her. One day, as I sat in my rocking chair nursing him, my mother called, angry about something. I don't remember what. But I remember this:

"I can't talk to you like this when I'm feeding the baby, Mom," I said. "It's important to be relaxed while breast-feeding."

"I'm drying up your milk!" she shrieked. "Is that what you're telling me? That I'm drying up your milk?"

And it was true. First one breast, then the other, dribbled to a complete stop.

It's hard to write about what happened next. When my son was six months old, he was diagnosed with a rare, rare disease—a disease so rare that the pediatrician had never even heard of it. A disease relegated to one paragraph in a medical student's textbook.

Seven in a million. That was the statistic. As my husband and I sat in the specialist's office, on the wrong side of that statistic, I held my son in my arms and felt something ice-cold and new, and yet bizarrely familiar, crash over me. *You see,* an evil voice whispered, *you don't get to be a mother after all. You overreached. What did you expect? That it was your God-given right to have a healthy child? To have a happy family life? That's for other people. You're going to lose every single thing that matters to you. You have no right.*

"You both look so gloomy," my mother said to my husband and me within days of our receiving the news. We were medicating our baby around the clock, watching him for any sign that things were getting worse.

"I always put on a happy face," she went on. "No matter what's going on. What's the point in wallowing in it? I don't know where you learned this kind of behavior."

"What are you talking about?" I asked her, as Jacob lay listlessly on the floor of her apartment. "You've never been through anything like this."

"Oh, please. I remember when I took you to the doctor for your eyes—you were . . . what . . . eight years old? And your left eye was weaker than your right? It was a terrible time."

"My *eye*?"

"And here you both are," she continued. "You're walking around with your long faces, dragging each other down. It worries me."

"About what?" I snapped, taking the bait.

"It's not good for your marriage, darling."

Those were the last words my mother said to me for a year. I cut her off with surgical precision, telling myself that I would deal with her again when and if my son was well. I had no choice, the way I saw it. I could either put all of my energy into saving my child, thereby saving myself, or I could allow my mother to remain in our lives so that she could destroy us all.

Would it be cruel to say that I barely gave her another thought? I was a mother lion, getting a crash course in two of life's most important lessons: how to be a good mother and how to live with the unbearable anxiety of not being able to protect my own child. I operated under an entirely unreasonable set of assumptions, holding on to them with tight fists and a clenched jaw: he *had* to be okay. That was all there was to it. My beautiful, fair-haired, blue-eyed boy, my sweetheart, love-of-my-life child. The alternative was unacceptable. Saving Jacob became my full-time job, as if, if I applied every bit of intelligence, will, determination, and financial resource available to me, I could save him.

One image stays with me from that year. The experimental medication that we had pinned our hopes on, the one narrow road out of that seven-in-a-million statistic, was a non-FDA-approved drug that we had FedExed to us each month from Canada. This drug came in the form of packets of white powder, which we had to divide into Jacob's five daily doses. We did this, my husband and I, by dumping the white powder onto a glass plate, and painstakingly cutting lines of the white powder with a razor blade.

I guess it's obvious where I'm going with this. For the better part of a year, that plate with the lines of white powder and razor blade was featured prominently on our kitchen table. Sometimes people would come over, whom we didn't know well, people who were unaware of what we were going through, and I'd see them eyeing the plate. And I'd remember another time in my life—a time when such a setup would have meant something completely different. Now, I held a tiny blue plastic spoon filled with strained plums and the miracle medicine, trembling, trying to get every last drop into my baby's mouth.

I will never know what saved my child. A talented doctor, the right medication, my own hypervigilance, my dead father watching over us, the prayers I whispered into the top of his head every night as I rocked him to sleep. Fate. Plain dumb luck. I will never know what saved him any more than I will know what stray dark cloud passed over him to begin with, threatening to engulf my little family—this family I hadn't even known I had wanted, but now could not possibly live without.

Then, once it was all over, once the doctor sent us home for the very last time, once I began to believe that we were all in one piece—my mother drifted back into my line of vision. She had been there all along, of course, hovering in the peripheral darkness. But now she was back. I remembered the promise I had made to myself, that I would contact her once Jacob was well again. I felt superstitious about it—like I had to keep all bargains I had made. It was one of the Ten Commandments, after all. *Honor thy Father and Mother*. And to top things off, it was September. Time for the Jewish High Holidays. Rosh Hashanah and Yom Kippur had always scared the shit out of me. The Book of Life was being opened in the heavens, and God was determining the fate of every human being—though this was unclear: did this apply to non-Jewish human beings as well? In any event, fates were sealed. Who was going to die, and by what method—all this was being decided by the big guy in the sky.

And so, I picked up the phone.

"Hi, Mom."

A cold, stony silence.

"Mom?"

"Hello, Dani."

She only used my name when she was incredibly pissed. As if calling me by my own name was the worst possible insult. And I felt it as an insult, a stone in a slingshot, released, hurling toward me. Her very tone of voice was frightening. I felt myself split in two, dividing: the part of me that had been strong for my husband and child, the part that had handled a hard situation with a fair amount of dignity—that part was still there. But it was shrinking rapidly, and in its place was a terrified, empty-headed, frozen little girl.

"I'm calling with some good news," I said, keeping my voice calm and even. "It looks like Jacob's going to be fine."

"Oh, thank God," she said. "You don't know how I've worried."

Why did it sound so false? Why did everything she said sound like she had rehearsed it in front of a mirror, trying her best to approximate maternal concern?

"I'm sorry you've been worried," I said.

"Right," her voice went back to its reedy coolness. "Like hell you are."

Time passed. The bullet we had dodged with Jacob gave me a strange, new freedom—the freedom to have as little to do with my mother as possible. Was she behaving even more horribly than she had for my entire life? I wasn't sure. But it didn't matter. I had lost my ability to tolerate her.

Occasionally, my husband and I would gather up all our defenses and bring Jacob to her Manhattan apartment for a visit. On our way over, I would feel something stir inside of me—a tiny ray of hope that this time it might be different—and I would try to kill the hope, squash it before it could squash me. I can still see her, swinging the door open with a flourish, her arms held out wide—not for a hug but in a gesture a diva might make during a curtain call. *Look at me!* she screamed from every pore. She wore T-shirts with big bold letters, or sweaters knitted with fuzzy angora animals. Anything to get Jacob to focus on her.

"It's Grandma!" She would get right in his face, blocking his path into her apartment. I would stand behind him, resisting the urge to scoop him up in my arms and flee. I squeezed my husband's hand tight. It was so painful to watch my mother try to interact with my son—I felt it as a physical sensation. I shrank, making myself as small as possible until I was a child again myself: lonely, afraid, and confused.

The more I withdrew from my mother, the more she attacked. I began to think of her as one of those unmanned drones the military uses in war zones. She was heartless, unstoppable. She didn't care what she was doing to us. It was all about having an impact—positive or negative, it didn't much matter. It would not be an exaggeration to say that she called a dozen times a day. When she was unable to reach me, she sent faxes, letters, FedEx packages. I walked around feeling cloudy and unhinged. Like I must be a dreadful person to cut off my own mother. Who could do such a thing? "You're destroying me," she said into my answering machine. "I have no reason to live." And one of her favorites, said so often over the years that it had almost lost its power: *Someday I'm going to die, and I feel sorry for you, because you are going to be very, very guilty*.

Understand this: I never stopped caring about her. I hesitate to use the word "love," but it's true. In a way, I did love her. She was my mother. She was inside me, a constant voice in my head. Sometimes, I thought her thoughts, and was unable to distinguish them from my own. I'd be driving, or taking a bath, or on the edge of sleep, and I would suddenly think, *I'm a terrible daughter. My mother is wonderful and misunderstood.*

In the meantime, every day I was learning how to be a mother myself, my feelings intensified by how close I had come to losing my son. The rest of my life—the part that had nothing to do with her—had gotten very good. My marriage had grown even stronger through the firestorm of Jacob's illness. Jacob was becoming an active, joyful toddler. Work was fine. And we had decided to leave the city and buy a house in the country. At least once a day, while pushing Jacob on the swing or watching him ride on my husband's shoulders, I had this thought: I could have missed out on this. I hadn't even known that I wanted motherhood. If not for that spontaneous, enormously lucky afternoon at the beach cottage when the veil of my fear lifted for just long enough, I might never have known. The pull away from motherhood had always been too powerful, a strong undertow drawing me out to sea.

The winter before Jacob turned four, my mother called me. I'm not sure exactly how long we had been out of touch, but it had been a while.

"I have these little things in my brain," she said. Was it my imagination, or did her voice sound weaker?

"What kind of little things?"

"Little things. Nodules." She paused. "Tumors."

It seemed impossible to me that my mother could die. I mean this in all seriousness. She was too tough, too huge a presence. I had long believed that she would outlive me. I was softer, more porous, vulnerable. She and I had been in a battle to the death for many years, and it was a battle I assumed she would win. I had pictured her at my own funeral, crying in the front row.

My mother had six months, on the outside. And so, I became her daughter once more. Of course, I did all the obvious things: doctors' appointments, nursing care, whatever she needed. The harder part was that we needed to be around her. After a lifetime of avoiding my mother, I was all she had left in the world. She let me take care of her. What choice did she have? If we had each swallowed a potent truth serum, our conversation would have gone like this: "You've killed me," she would have said. After all, she had said as much in the past. "Better you than me," I would have responded. And I would have meant it. I was younger. I had a small child. I had a right to live. Didn't I?

But we never had such a conversation. When well-meaning friends tried to tell me that my mother's illness was a blessing, because now we might have time to make peace with each other, I smiled and nodded—they were only trying to help—but I knew that we had long since passed the point where any sort of peace might be possible.

"Jacob, it's Grandma!" She still opened the door with a flourish, but the fight had gone out of her. She was bald from radiation, and leaning on a cane. Gone were the jaunty sweaters. She grew smaller and weaker, until, for a very brief period of time—a week, maybe two—she became a person it was easy to be around. The anger had vanished. She drank in the world that she was leaving. A day stands out in my mind: a sunny, gorgeous, early spring day, the last time she was able to be driven to our house in the country for a visit. We somehow managed to get her outside and into a chair, and I sat next to her as we watched Jacob play on his jungle gym. We just sat there, my mother and I, both of us enjoying the moment. We had never, in our forty years of knowing one another, been able to have such a moment. I had hated being her daughter and—this struck me with all the force of a physical blow: she had hated being my mother.

She sat on the green expanse of lawn behind our house, the view sweeping miles into the distance. Jacob shrieked and laughed, kicking a ball downhill. My mother simply watched him with pleasure. Her eyes were not flickering all around, looking for ways to disapprove, to judge, or to make herself important. She was not maniacally working to place herself at the center of the universe. She was simply existing, and I was existing alongside her. It was the saddest I have ever been.

LAUREN SLATER

Second Time Around

IN MAY OF 2001, I wrote an essay for a national magazine, explaining why I would only have one child. In this essay, I described my rather wrenching decision to abort my second pregnancy, a decision based on solid common sense: lack of finances, lack of time, lack of personal and emotional resources, a marriage already strained to its breaking point, a commitment to career. Now, years later, from where I sit high in the house's attic, typing up these words, I can hear the caterwauling of my second-born, produced from a third pregnancy. At six months old, he has a head covered with the fine nap of blond fuzz, and hands that look like a sage's. His name is Lucas. He should not be here. He should be wherever it is the prebirthed people live, in a sea of swarming atoms, or high, high up in the air, near Jupiter.

Despite all the common sense, the careful thought, the agonizing decision, the definitiveness of the door closed on whether or not to have another child—despite it all, I went against my better judgment, and two years after the abortion I did it. I did it just as I rounded the bend into my fortieth year. I did it as my periods started to lighten and the skin looked crumpled around my eyes. I did it after I bit into an apple one summer morning and heard the sharp crack of a splitting tooth; I did it in decline. I did it because of decline. I did it holding close to myself the image of me and my husband as old, old people (if we are lucky enough to get that far), and the thought of who would be there for my daughter, when we died. I did it as a strike against death but more important than that, I did it because, in the very end, having one, for me, was just too risky.

I did not want to have my first child. Before she came to me, and before I came to love her, I dreaded the thought of motherhood, all those hours spent on the playground, or in Chuck E. Cheese. I had heard women talk about "baby lust" and knew I possessed not a drip, not a drop, the drive toward procreation almost absolutely absent in me. Motherhood went against my nature, which is brooding and acerbic and self-consumed. My husband wanted our first. I did it for him.

After forty-eight hours of grueling labor, I had my firstborn, Clara Eve, in 1999. I don't think I loved her right away. She was a truly beautiful baby, with interesting cheekbones and a mouth like a little red bow. She slept deeply. I kept thinking *she's mine she's mine she's mine*, but she didn't feel like mine. She didn't feel like someone else's, either. She felt otherworldly, as though she were surrounded by stars.

I recalled, during those days of recovery in the hospital, that there had been a study, or maybe several studies, showing how mothers instinctively recognized their babies by smell. If you put a new mother in a roomful of babies, according to the study, her nose would take her lickety-split to her scented progeny, and vice versa for babies. If you gave a newborn baby milk-soaked rags from her mother and milk-soaked rags from another mother, the baby always put her mouth right up to Mom, closing around the cloth, drawing in and down. These studies haunted me. While still recovering from my C-section in the hospital, I would go into the nursery, where troops of babies, all in blue-striped hats knotted at the top like gnomes, slept in plastic bassinets. I remember walking from crib to crib, trying to guess, or smell, which baby was mine. I purposely did not look at the nametags. All the babies seemed exactly the same to me. All of their mouths were adorable. All slept intensely, as though dreaming of the place they had left. My heart would quicken there in that nursery, and I would peer and peer and, at last, pick a baby, the one I believed to be mine. I played this horrid little game many times over and over. I picked my baby maybe one quarter of the time, and the rest of the times I picked someone else's baby, and it was with dismay that I would focus my eyes on the bassinet nametags—Jack, Annabelle, Galileo. Where was my girl, Clara Eve? Oh, there she was! Two rows over. Either the nurses had washed her fragrance away, or there was something seriously wrong with my nose.

After three days in the hospital, I took her home. It was summer, hot hot hot, light glaring in the nursery. Over the weeks, my scar turned into a neat, pink seam, a four-inch slash, her tiny exit door. Time passed and clocks ticked, and days folded into nights spent rocking and feeding, and her cries were singular, the only ones I heard, and she started to smile, and she started to haul herself up by holding the edges of things, and I came to love her. I fell in love with my baby girl when she was just about eight months old, one month after I'd aborted her sibling; I fell hopelessly, horribly, dangerously in love. She was reaching up to grasp the rims of whatever she could find, pulling herself unsteadily to her tiny feet, and I saw, then, I saw with total clarity how from this point on her world would be full of sharp points and hard floors and so many dangerous angles! This sense, that my child is in danger, has not abated over the years. Perhaps this is what mother love is. Perhaps I have finally entered into instinct. The older my child gets (she is five now) the more ferociously I worry about her, and there are sharp shards everywhere. The one thing I absolutely could not bear would be to lose her, or have her hurt. And precisely because this is so unimaginably horrible to me, I spend an obsessive amount of time trying to imagine it. The world is now a series of distinct, personal flesh threats. When the children were taken hostage and then killed, in Beslan, in the early autumn of 2004, I spent as much time trying to imagine what it was like for the parents to suffer such a total loss as I did imagining the terror inside that gym, where bombs were put in basketball hoops.

I, in part, attribute my decision to have a second child to terrorism. When I wrote that magazine essay, when I had that first abortion, the planes had not been taken. We didn't know what the world was. We were blind. Then we saw our blindness. It's not only that I'm afraid my child will be the victim of a terrorist act—though, of course, I worry about that—it's more the sense that I now see the world stripped of its pretense. I see that we are all still primitives living on Pleistocene plains; I see that we are slaves to rage and greed. I see we have weapons, that we are tool builders, that we hunt and are hunted, and that even sleeping in trees will not save you. While I've always "known" this, it became sharply obvious to me on September 11, and its obviousness grew with time. Having a first child was not an instinctive act for me, just its opposite, but having a second, in the end, I had to do it. I did it out of raw, primitive terror and the age-old threat of loss. My own hypothesis—utterly untested—is this: the drive to have a first baby varies from person to person, but the drive to have a second, as a means of protecting yourself against loss and your child against loneliness, that is where pure biology begins to play its role. If you have a first, not having a second will always seem sad and dangerous. But if you never have a first to begin with, then you're free and clear.

When I got pregnant with Lucas, I at first tried to talk myself out of it. I tried to think it through rationally, although rationality is not a strength of mine. I thought about overpopulation and food sources and global warming. I thought that this was not my second pregnancy; it was my third; the second one I had aborted, and what was different now? The truth is, everything was different. In between the abortion and Lucas, I had come to know mother love, and come to understand that the world around me was gobbling up its citizens with ferocious speed. These two facts made my initial qualms seem almost irrelevant. We didn't have the money, fine. My marriage was already stressed. Too bad. Let's go. I tried to get pregnant; I did it on purpose and it worked right away. We were blowing up Afghanistan and Iraq. While, all over the globe, bombs were falling and people were ripped up, I did what the human race does in such situations. I had sex. Two weeks later, I began throwing up. My husband and my doctor said it was all in my head. No one got nauseous so soon. But I did. I didn't even have to take a test. I knew I was having my second. I was sick for eighteen weeks, retching, dehydrated, anemic, and not once did I seriously think of aborting it. I ate only watermelon.

This is not to say I wasn't ambivalent. I was. It's just that the ambivalence did not have the strength of the fear that propelled me forward. Nevertheless, especially during the eighteen weeks of vomiting, I thought, *what have I done? What am I doing?* One night, my husband said to me, "We were happy with one; we don't know if we'll be happy with two; we don't know if we're strengthening or weakening our family." He said this to me just as we were getting ready for bed. Our room is blue-violet in color. It has a single skylight cut in its sloped ceiling, and on clear nights, you can see planes swimming like sharks across the sky, and the javelin sharpness of stars. "We might not be happy with two," I said, while above me a jet glided in its ascent. "But," I said, and then I couldn't think of how to go on. We turned out the light. *Happy happy happy*, I thought. We live near an air force base. The fighter planes are trim and dart around. They lean in on one wing, like a gymnast showing off her skills. They occasionally nose down and then zoom up and then disappear, leaving behind only their echoing roar. *Happy happy happy*, I thought as the fighter planes went by. I couldn't sleep. The baby was hanging off my heart. It hurt. I understood then that happiness was not a primary drive. Our constitution got it wrong. We didn't want to be happy. We wanted to extend.

I grew enormous with my second child. I gained eighty-eight pounds. It was disgusting. It was unreasonable. I swear, for the first eighteen weeks, I ate only watermelon. But it was as though my body knew to plow forward, to pack on protective fat, and by the thirtieth week, even my face was huge; I looked nothing like myself. My husband and I went to see a lawyer. We went to set up wills. We hadn't thought to do this with our first, but now, it was as though we'd deepened into another level of parenthood; we couldn't fake it anymore. With one, you can retain aspects of a child-free life. With two, that becomes impossible. You are sucked into a stream of Graco and Huggies and ant farms. So, we set up a will. I was eight months pregnant and the lawyer peered at me suspiciously from his desk chair. "Who will be the guardian?" he said. We thought and thought about that. At last, we decided on our babysitter. Two weeks later, the wills came in the mail. I opened them up. It was a little like reading my own death certificate; I could smell death everywhere. The babysitter was thrilled.

The night I received my will, I lay with my daughter on her bed, in her yellow room, my bulging belly pressed against her back. *I am lying with my two babies,* I thought, and then I fell asleep.

On January 20, 2004, Lucas was born via C-section. They carved him out of me and held him up, blood-speckled and snuffling. He was just about ten pounds, full of himself, secure in the knowledge that he belonged here—right from the start. We brought him home on the coldest day of the year, when the air was cracking and painful, the trees black. My husband said, as we were in the car, the two babies in back, "Well, now we're really a two-kid family," and I could hear several things in his voice: happiness, hesitancy, fear. At every rut in the road, my wound sparkled with pain. I was high on Dilaudid and had that postpartum-weepiness mood, a mood that amplifies the meaning of everything, so you are at once full of rapture—I had a son! I had a second child!—and also full of despair.

Eventually, the vicissitudes of my mood evened out. Slowly, I got to know Lucas. My daughter displayed zero jealousy, which made me think she'd shoot up a playground someday. The winter eased, and the air began to smell like spring. Walking with Lucas in his sling outside, stepping over the wet spots in the road, my cuffs splattered with mud from speeding cars, I thought of what people had said to me regarding my reason for wanting a second. "You won't ease your worry," they'd said. "You will only double it. With a second, you'll have two to possibly lose."

And in some ways, these rational, thoughtful friends and family were right. Lucas came down with a strange illness; his throat was full of white spots. He screamed and screamed. Worrier that I am, I immediately assumed the worst. Smallpox. Measles. Death. These fears were in no way easier to bear because I had my girl. No. I took him to the doctor, and they diagnosed him with hoof-and-mouth disease.

He's better now. At night, I feed him and I can feel the fontanel hardening, the plates of bone growing together in his skull. Perhaps because my daughter had primed me, I fell for him more quickly than I did my first, say around week three. I feed him, and I press my lips to the fading soft spot and I see the place in his head where his pulse bubbles up. My boy. I hold him close, and it is never close enough.

While in some ways I've doubled my worry, I feel I have also eased it in a way too primitive for words. Let me be clear; if I lost either one of my kids, especially if they were hurt in the process, it would be devastation for me. But now I know I would have to live through it, for the other one. And when I see Clara playing with her brother, I think it was worth it, because she truly is less alone in the world. Yes, we have less money. Yes, we cannot afford a house in a nicer neighborhood, or a private school; yes, I have lost whatever remnants of my child-free life I once had. I have become exactly the kind of woman I said I would never be. With two, in love with two, responsible for two, I have at long last set my beloved friends aside. I have no social life. But somehow the world feels a little bit more right to me. A gap has been closed. My girl has another. My eggs are not in the same basket; I have two eggs, and they're each in their own separate spots, hatching, growing up, all around them the dangerous, smoking world. I hold my children's hands. I did not know it was possible to love so thickly. My life has in some ways been narrowed by sheer fear and flailing instinct. I have less and I have more. What matters most is this: I have absolutely no regrets.

PETER NICHOLS

We'll Always Have Paris

When I was twenty-nine, my wife became pregnant. We were broke and living in the Caribbean aboard a small wooden sailboat. We were planning to sail across the Atlantic to the Mediterranean that year, and then we were going to build a boat and sail around the world. It wasn't the right time for us to have a baby.

We had less discussion about the abortion than we usually had about where to drop the anchor at the end of a day's sail. We didn't like the idea, but the choice seemed clear to both of us, and we were glad we had one. Yet after the procedure, we crept back to our boat and found inside it an emptiness we hadn't remembered. While my wife cried in her bunk and I made tea for her, I couldn't stop thinking about who it might have been we had removed from the world that afternoon. For comfort, we told ourselves there would be plenty of time for children in the future.

When the future came, I was living in Los Angeles, writing screenplays, and married to someone else. As we neared the end of our thirties, my second wife and I figured it was now or never if we were to have children. With that biological imperative—rather than a clearly formulated desire for a child in our tidy house or in our busy lives—we began to try to make one.

Six months later, we visited a fertility clinic in Beverly Hills. Everything seemed in order with us: my wife made eggs of apparent quality; and I was able to produce the requisite hundreds of millions of sperm to expect that a handful might survive the journey to reach my wife's eggs. But time was not on our side, the doctor said, and to expedite things, I began to inject my wife regularly with fertility drugs. On the appropriate day, my sperm, launched into small drinking cups, was washed by the fertility lab and introduced into my wife's uterus.

We had by then begun to imagine and anticipate life with a child. We discussed names. We browsed the family shelves in bookstores. We noticed children everywhere. The world was overflowing with them, they seemed so readily available; where was ours? We began to feel that we would be cheated of one of life's essential experiences if we didn't have one of our own. Our lives weren't empty or without fulfillment, but they had become so calculable, even the unforeseen could be guessed at. This—a greediness for a deeper experience—was what drove us, rather than a yearning for the cute little people we saw everywhere. We didn't fuss over other people's kids, or try to spend time with them. Parents, from where we stood, were boringly self-absorbed in their children and their domestic lives. They talked about nothing else, their houses were messy. Yet there must be more to it than met the eye, and we had become determined to understand this unknown for ourselves.

But my wife did not become pregnant. While we waited, and felt mounting disappointment month after month, our future by ourselves, without a child, began to look increasingly barren. With nothing to open it up or carry us further, a terrible staleness yawned. It seems obvious to me now that we began to feel, if not to admit, that by ourselves we were not enough for each other, that we were hoping a child would make a difference to our marriage. Following no success at the lower levels of medical intervention, the process of in vitro fertilization loomed. "Let's take a break," we said, but it was the end of trying, and it marked the end of all our trying together. A short time later we broke up.

My screenwriting career seemed almost as fruitless, and Los Angeles became for me an unbearable taunting panorama of failure. The sudden flowering careers of others, the beautiful houses filled with children. I moved away to a rented pool shack in northern California. I was in my mid-forties, broke again, living like a student, with a bag of clothes, boxes of books, an old U.S. Mail Jeep, a bicycle, and running shoes. It was as well, I thought then, that I had none of the children that might have come my way: the fifteen-year-old, had it not been aborted, from my first marriage, whose age I noted as each year passed, or the twins or triplets our fertility experiments might have brought my second wife and me.

With no one to please except myself, I began writing, with almost no idea of shape or outcome or hope of publication, what would become my first book. It was an account of the years I spent living aboard that wooden boat with my first wife. One November morning, after a long run in the rain on the dirt fire roads along the ridges of Mount Tamalpais, I came back to my shack and found a message on my answering machine from an agent in New York. He'd read the first seventy pages and said he thought he could sell it. And promptly he did.

After that, in the space of four years, I wrote and published a novel and two other books of non-fiction.

I arrived at fifty unmarried, and, for the first time in my life, I was modestly successful. I had some money, I was single and uninvolved, I still owned almost nothing. I was as unencumbered as a monk. I was also lonely and hoped for something that might end that: a relationship—and, who knew, maybe even kids, but for that to happen I'd have to meet and fall in love with a woman in her thirties who would fall in love with a man in his fifties, and we'd have to be together long enough to believe that it would last . . . it seemed less and less likely, and I felt myself adjusting to that. I'd be grateful to find love and companionship. Meanwhile, I was enjoying a rare freedom that, for one reason or another, I suspected could not last, so I decided I'd indulge a long-held fantasy: to live in Paris and write a book there. To drink coffee and read the *Herald Tribune* every morning at my neighborhood café, write through the mornings, walk the beautiful streets and get to know Paris in the afternoons, better my French, learn more about wine and food. And perhaps, *naturellement,* meet one of those pouty, smudged-looking heroines from the French New Wave movies who have asymmetrical faces and are perpetually ready to pull off their chic sweaters. Françoise, Veronique, Brigitte, Loulou . . . she'd probably be a smoker, but I was ready for anything.

I rented an apartment in the fifth arrondissement, on the Left Bank, near the Sorbonne and several good street-markets. My café turned out to be the Rostand, on Place Edmund Rostand, a lovely, unusually spacious belle époque salon of potted plants, paintings, wicker chairs, marble tabletops, with a view, through the tall glass windows, of the Luxembourg Gardens across the street. I went there every morning for my coffee and newspaper, and again many afternoons for the pleasure of the place, to read or scribble notes, and to gaze at its attractive clientele. I became a regular, not a tourist, soon recognized as such by Olivier and Sophie, the poised, smartly dressed, punctiliously welcoming hosts who greeted the café's customers. In the winter, I sat inside with a *vin chaud*, beside the wood fire; in warm weather, *bière pression* or *citron pressé* on the sidewalk. I was writing a book for which I'd been paid a large advance. My life felt, if not complete, unprecedentedly rosy.

Five months after moving to Paris, I was sitting inside the Café Rostand on a January morning, with a friend who was in Paris, researching a magazine story, when my cell phone rang. An American woman's voice said, "Peter? Hi, you don't know me but I'm a friend of" a magazine editor for whom I was writing a story. The unknown caller had just moved to Paris, also to write, and friends had given her a list of people to call. I was one of many.

It wasn't to be Françoise, Veronique, Brigitte, or Loulou, then, but Roberta from Texas, by way of Santa Fe. Not a New Wave heroine; more 1940s Hollywood, a tall, whipsaw-smart, gloriously American girl in aged cowboy boots and a vintage Yves Saint Laurent coat. I fell crazy in love with her. I courted her through the spring in Paris, which was even more romantic than it sounds. The city's flower shops, *chocolatiers, confisseries,* markets, bars, bistros were suddenly staffed by angels who were collaborators in our romance. Paris loves lovers, and makes a playground for them like no other place on earth.

After a six-month courtship, we went to Santa Fe, where, in front of Roberta's variously amazed or dismayed friends and family, we were married.

Back in Paris, we continued to live a bedazzled, wonderfully carefree existence. We found a large, light, high-ceilinged, tall-windowed, two-bedroom, two-bathroom apartment on rue Saint Jacques—even closer to the Café Rostand than my former pied-à-terre—which we furnished from trips to the flea market at Clignancourt. We wrote in the mornings; afternoons we rendezvoused at Gerard Mulot's patisserie or by the carousel in the Luxembourg Gardens; we traveled by first-class trains to see friends and family in Provence, Rome, London, and Spain. Our life together seemed at last to be the one each of us had been destined, after long delays and much tribulation, to live. We had died and gone to Paris.

Roberta turned thirty-nine the day before our wedding. She had never been married before and had almost given up on the idea of having children. Now, for both of us, the clock was ticking. We looked at each other and imagined the child we might make together: it would be more of us. It was too heady an alchemy to resist. Certainly, we would have liked more years together without children. We talked about it: what would we do, go trekking in Patagonia? Stay in Paris or try Prague? But we had both traveled enough. If we were to have our own children, we had to start now.

In order not to waste what might be precious months of diminishing possibility, we went for tests. Roberta produced numbers indicating the fecundity of a maiden, and, on a rainy autumn afternoon in a Parisian laboratory bathroom strewn with French porn magazines, I was able to deliver, still, an abundant sufficiency of suitors. On a Tuesday in late November of 2002, Roberta had an ultrasound examination at the American Hospital in Paris for the purpose of checking the readiness of her eggs and fine-tuning the timing of our monthly effort.

"Try Thursday," said the doctor.

Thursday was Thanksgiving, a cold, rainy, late-autumnal day in Paris. We made love in the afternoon, and then went out to have Thanksgiving dinner with French and American friends.

In mid-December, Roberta's store-bought pregnancy test produced a positive result. Like two kids conducting a very grown-up chemistry test—which felt at moments something like making a homemade bomb—we were amazed that it had actually worked. And a little stunned. Our brave new world was a few months away.

Through the winter, as we continued to savor the dwindling days of our Parisian lifestyle, came unsettling considerations. Our lovely apartment, our love nest, our first home, was a walk-up on the fifth floor. With no elevators for strollers, it became clear to us that we would have to move. We began meeting with French physicians, and for the first time, we were confronted not with conspiring angels but peremptory and decidedly patriarchal Gallic officialdom—the cold, rude-seeming dismissiveness for which the French, Parisians in particular, are famous. The doctors we consulted aimed their brief comments, even their answers to Roberta's questions, at me, as if she were my twelve-year-old daughter with tonsillitis. This didn't sit well with Texas. The hospitals and offices where we met them were old and dreary. We read that physicians and anesthesiologists often went on strike. The labyrinthine and obfuscatory nature of French bureaucracy almost precludes any possibility of suing a French doctor, so they don't bother to pretend they are not in a hurry and have all the time in the world—and interest—to listen to all your silly questions. French doctors are not "there for you." They are more often than not visibly harried, or bored, peremptory, quickly dismissive, and superior. "What's the difference between God and a doctor?" goes a French joke: "God doesn't think He's a doctor." Roberta's experience of the amniocentesis procedure at the hands of a French OB/GYN was cartoonlike: "Prepare for the puncture," he said, and raised a gigantic syringe above her swelling belly and stabbed her.

This maggot of discontent infected the rest of our francophilia. The French do, I feel, have their priorities right about many things, but this means that shopping in Paris for everyday items such as shampoo or Sheetrock screws requires protracted expeditions and lacks the charm and ease of obtaining any of three hundred varieties of exquisite cheese. We found ourselves missing the availability and speed of acquiring anything in America, and the solicitous consideration of American doctors for their patients. The sense of separateness we'd enjoyed as foreigners began to loom as a disconnectedness from our surroundings. We wanted, when our baby came, to belong somewhere. By resistant but ineluctable degrees, we decided finally to return to the United States. We would go "home" to have the baby, but we wouldn't give up Europe, we promised ourselves. We would come back.

In the last trimester of Roberta's pregnancy, we moved to a seacoast town in Maine. We'd decided upon the place almost by throwing a dart at the map. I like Maine, which feels like a slower, older, less spoiled appendage to the rest of the United States. I'd taken Roberta there for a few days while I was on a book tour, before we were married. She liked it, too. "Sure, let's go to Maine," she said, just like that.

Small-town America, which may seem the acme of dull ordinariness to most Americans, looked awfully good to us after Paris. After you've battled long enough with the intentionally obtuse and intractable proprietors of that city's immaculate boutique pharmacies, a Rite-Aid drugstore on Main Street is heaven. The doctors and staff in the local, ultramodern OB-GYN unit were agreeably sympathetic and filled us with confidence.

We decided to rent for a year, to see how we liked our new town, but within six weeks of our arrival we impulsively bought an old Victorian house and began extensive renovations. Later, wondering what had gripped us, we read of the manic power of the nesting instincts of parents-to-be.

Our baby boy was born by cesarean section after a long day of difficult labor on September 1, 2003—Labor Day. Augustus Paris Nichols we named him, after Roberta's maternal grandfather and after the city in which he was conceived that had brought all three of us together.

We've had fall, winter, spring, and two summers as parents in Maine. At one year old, Gus sometimes sleeps without waking through the night. What rare nights those have been for us. We're coming out of a long tunnel of nocturnal desperation. Until recently, I felt more sleep-deprived than when I was crossing the Atlantic alone in my sailboat many years ago, when I slept no more than half an hour at a time for weeks on end. Everything you hear is unimaginably true:

you become more tired than you can believe; you make do with less sleep than you thought possible; with this exhaustion, sex takes a distant, if discontented, backseat. Sometimes Roberta and I are so tired we can't even speak to each other, apart from passing on essential information: "His diaper's just been changed."

In the beginning, having a baby is like finding a seal pup on the beach. It's helpless. It wriggles, shits, and cries in the most plaintive way. You can only clean it, wrap it up, and endlessly feed it. There is no time for any other effective occupation. This is perhaps the greatest shock: the complete shutdown of all the rest of your life. You vie with your spouse, sometimes acrimoniously, for time, the lost treasure. Everything changes and so will this, you tell yourself, and so constantly do the parents of older children. "Enjoy this precious time!" they tell you, "it goes so fast!"

For longtime free agents, like Roberta and myself, who have had and spent our own time like millionaires, this is a blindsiding blow that no amount of reading, anticipation, warning, or even the long confinement of pregnancy—which, with sleeplessness and a gradual closing of worldly doors, nature cunningly designs to mimic the conditions after birth—can prepare you for. It amazes me to remember the free time I once had: the hours I used to spend reading or doodling in the Café Rostand, whiling away an afternoon before going to a movie. Shopping for and preparing dinner with such doting deliberation. The trips to London and Rome. Time in which to write, not to mention those self-indulgent disciplines of the time-privileged: walking, running, working out. These things are part of a former life. People tell us such times will come again. "Enjoy this precious time!" they say again. We'll try, but when will it be over, please?

"We'll always have Paris," Roberta and I say to each other with grim humor. Now we most certainly mourn our lost episode of love, freedom, and romance in the city of light. It was far, far too brief. We could have wolfed down years of it. What has overtaken us sometimes feels like a mistake: the wrong reality, one that belongs to someone else. At forty and fifty-three, when Gus was born, Roberta and I had reached what passes for middle age with only ourselves to look after. We were profoundly accustomed to being alone. In this late arrival at parenthood lies the degree of difficulty and adjustment we have felt in the last year.

But also the joy, for there is Gus in the place of everything gone. From the moment of his first breath, there was something else, which has given us infinitely more than we have lost. As is commonly said, it's something you cannot know until you experience it. When I first saw my son, gooey from his uterine fastness, saw what we had made from all that love in Paris, I recognized him, physically, as I would recognize my own foot or hand, but in another way, too: he looked like the promise of all that Roberta and I had felt together, fulfilled.

"Please see where he goes," Roberta pleaded dopily from the operating table, with the parent's second fear for their child—after first seeing that it is as perfect as God and nature meant it to be—that it will be lost or confused with another baby. I would have recognized our boy and been able to pick him out in a nursery filled with five hundred other babies, though he was well labeled and we were the only action in the maternity unit on that Labor Day.

Since his arrival, through all the exhaustion and preoccupation, I have grown another part of myself: another whole heart that dwarfs my preexisting organ, this one the size of sperm whale's, a brand-new, gargantuan muscle, developed by strange and powerful paternal steroids. It will beat inside me until I die. The earth has tilted on its axis and is a different place. I dream now of what my little boy and I might do together in the years ahead. When he cries at night and I pick him up, I think of my own, long-dead father and can feel now what he felt as he carried me through sleepless nights.

The seal pup has now been replaced by a sentient little person whose babyhood is passing with dismaying speed. Gus now walks and runs. He understands much of what we say to him. He talks, even, saying "Mama" and "Dada," but he understands and conveys far more. I can see already the little boy he is becoming. When he sees me, his face splits into a huge smile that now contains six teeth. When I squat and call his name, he laughs like a boy on a roller coaster and runs to me and throws his arms around my neck. There is no other connection on earth like this. "Little Gus," I say to him as I lift him up, and his fists clutch my hair and I feel his breath on my neck, and stroke his ten thousand-thread-count skin. My most compulsive addiction is to the sound of his laugh, and he laughs often and readily and is filled with an uncomplicated joy of life.

I can still hardly believe it, and I don't think I'll ever get used to it. This so nearly was not mine, and I knew it. I've been traveling recently, spending a week at a time away from home, and then I feel the sense memory of my old life alone, and the thin insubstantiality of it. When I come home and see Roberta and Gus, and watch his face split into his pure, toothy grin and hear him yell as he catches first sight of me, I'm aware that I've been granted access to a new plane of existence, one I could not have imagined, and would not now live without.

MAGGIE JONES

Surprise, Baby

When my daughter was born, I was three time zones and six states away, reading the newspaper in my parents' kitchen. For more than four years, I had tried to conceive her. I had injected myself with hormones and swallowed hundreds of pills. I had allowed doctors to shoot dye into my fallopian tubes and to biopsy my eggs for chromosomal abnormalities. Then, on a Monday evening in June, the phone rang. "A baby has been born," the voice on the other end said.

But before that moment happened—before it *could* happen—I'd made very different plans. I was in my late thirties, recently married, happily immersed in work that held my attention until late in the evenings. Though I wanted to become a mother, I wasn't in an enormous hurry. I was conflicted about the sacrifices, and as the daughter of generations of fertile, healthy, strong women, I had the hubris to believe that pregnancy would come easily for me.

In fact, I did get pregnant quickly the first time. But something about it never seemed right—I didn't feel pregnant. Six weeks later, during a snowstorm, I started bleeding and spent the weekend in bed, nursing my pain with plenty of codeine and mourning a baby I realized I did very much want. When I conceived again several months later, I was so nauseated and exhausted that I dared to whine to my husband "I hate being pregnant," and in those moments, I did. But I loved being pregnant the day my doctor pulled the ultrasound wand across my stomach and I saw that flicker of a heartbeat on the murky black-and-white screen. "It's going to work," I said, leaning into my husband as we left the doctor's office. At home, I tucked the ultrasound photo in a drawer and indulged in pregnancy Web sites: I learned the due date (a fall baby); I weighed the pros and cons of prenatal tests; I tracked the week-by-week progress of our child.

At my next ultrasound, my doctor stared at that same black-and-white screen for a few moments too long, and I knew: The heartbeat had vanished. "I don't understand," my husband whispered from the foot of the examining table. That evening at home, the ultrasound picture still sat in my drawer; all those Internet images of fetal hearts, spinal cords, eyes swum in my mind. The details were not only useless now; they felt cruel.

My next pregnancy failed, too. As did my next. Though several family members and friends offered tender words of sympathy, others seemed anxious to find a silver lining. "At least you get pregnant," one friend said. "It's probably for the best—something was wrong with the baby," said another. By the fourth miscarriage, I simply stopped telling people. What was there left to say about so much loss?

When my mother gave birth to six children in the 1950s and 1960s, she never had ultrasounds or two-by-three-inch snapshots of her fetuses. She didn't use home pregnancy tests or ovulation kits. If she had very early miscarriages, she probably never knew it, never knew she'd lost a baby.

I, on the other hand, was skittishly alert to each pregnancy, and to each loss. At night, in the early mornings, during hours when I should have been working, I searched the Internet for answers about recurring miscarriages and infertility. I consulted with various doctors, but none of them could pinpoint a problem. I visited acupuncturists, one of whom suggested I was ambivalent about being pregnant, another that I needed to practice "holding on" to things. I saw a nutritionist, who recommended I cut out the wheat and increase the protein. In the process, I went from being casual about pregnancy to becoming a ravenous collector of information, always believing that the answer was out there if I only tried a little harder and dug a little deeper.

By the time I signed up for in vitro fertilization, I wasn't an easy or a happy patient. The process seemed the coldest in all of fertility medicine. Not only is sex unnecessary, but IVF leaves little if any room for spontaneity or mystery, the very things that I'd always assumed would be part of creating a baby.

After two weeks, forty-two injections of fertility drugs, and every-other-day trips to the clinic, my doctor retrieved more than half a dozen eggs from my body, and the lab tech mixed them with my husband's sperm in a petri dish. Several days later, on the morning my doctor would transfer my embryos to my womb, he handed me a photo of my three "best" candidates—the ones that had earned a grade of "A" from the embryologist. I could be aggressive, he said, by implanting three or four embryos, or conservative and put in two. "Talk it over," he said, before sending my husband and me to a nearby diner for a few minutes. It was around eight-thirty a.m. and people were buying coffee and bagels and muffins before heading to work. I watched them with envy: Did they have children? How did they conceive them? Did they have to make the kind of choice I was about to make? Under the small table, my knees were touching my husband's. He looked at me and shrugged. "You have to carry the babies; it's up to you," he said. The more embryos we transferred, the greater the chance of twins and triplets—and the greater the risk of prematurity, illness, and disability. Yet if I put in only two and neither survived, would I wonder if the embryo I'd left behind was the healthy one? On the table, in between my husband and me, lay the photo. I looked at it again—it was of what exactly? Clusters of cells? My future children? I closed my eyes for a moment and chose three.

Two weeks later, when the doctor telephoned, my husband and I each picked up a separate phone, so we could hear the news together.

The pregnancy test was back and it was negative. I wasn't even tentatively pregnant as I'd been so often on my own.

But when one procedure failed, my doctor invariably had another one to offer up. This time, he recommended IVF in conjunction with "preimplantation genetic diagnosis," a test for common chromosomal problems (Down syndrome, Trisomy 13, Trisomy 18) that entailed drilling through the shell of each of my eggs, removing a cell, and biopsying it. Even if the procedure didn't damage my embryo, it might prevent it from attaching to my uterus; in trying to create a pregnancy, I could ruin the chance of one. It scared me. And yet I'd come this far, invested so much financially and emotionally. I didn't want to say yes exactly, but I wasn't ready to say no.

After two weeks of the usual IVF routine, eight of my microscopic embryos underwent their biopsy; because of my age, the doctor expected that only about half of my embryos would be free of the chromosomal glitches that led to miscarriage and contributed to infertility. And to give the healthy ones the best chance of sticking, they'd have to be transferred to my womb as soon as possible. The next morning, the test results were supposed to be in, but I hadn't heard from the clinic. I called the nurse. I called the doctor. I paced the living room like a protective mother awaiting news of her children. An hour before the scheduled embryo transfer, there was still no word, and we had a forty-five-minute drive to the hospital. My husband prodded me toward the car. When we were just a few minutes from the clinic, my cell phone finally rang. "You have one normal embryo," my doctor said.

In the hospital, a slightly distracted doctor I'd never met pulled her chair close to mine and removed a slip of paper from the pocket of her blue scrubs. It was folded in half, so we could see only one side, on which she'd written the number of cells of each of our embryos. On the morning of the biopsy, we'd had eight healthy-looking embryos, most were at least eight cells each. "This is just what we like to see," the doctor said. My eyes were bloodshot from crying in the parking lot and the doctor was rambling about my lovely embryos, when she and I both knew their surface appearance was irrelevant.

"Just tell me about the biopsies," I blurted. Then she flipped to the other side of the paper, where she'd scribbled Trisomy 21, Trisomy 22, Trisomy 16. Only embryo No. 3 had tested completely normal, but it was a mere five cells. Now our hopes would have to be invested in this—the slowest, the tiniest, the most vulnerable of the lot.

In the days following, I lay on the couch at home and imagined No. 3, smaller than a pinhead, suspended in my womb. Sometimes I would envision it attaching to the wall of my uterus, getting nourished by my body and thriving. Other times I pictured it floating, lost, like an astronaut untethered from his spaceship. For some reason, I thought it was a boy. Small, but quick and spirited. The little boy who could.

But two weeks later, my pregnancy test revealed that No. 3 didn't survive. Maybe it was me—maybe I was too stressed. Maybe the embryo wasn't strong enough. "You might have two or three healthy embryos the next time," my doctor reassured us. Or we could have none.

I thought of two friends who'd stopped far short of what I had done. Both had adopted baby girls and were as happy as any mothers I knew. So, what about me? Four miscarriages, two failed IVFs, three years. Why couldn't I let go? I could have motherhood, after all, through adoption. And it was motherhood I wanted, so much more than I wanted to experience being pregnant. But I was like a gambler at the roulette table, whose number is always about to hit.

The problem was, I wasn't only missing out on time to be a mother. The repeated miscarriages and unsuccessful IVFs, so much effort followed by so much failure, was diminishing me in some way. In my dogged pursuit of fertility medicine, I was losing faith that life could unfold in more interesting ways than I could plan for.

"A baby has been born." It was my husband's voice on the phone. He was in Los Angeles, where we lived, and I was visiting my family in Kentucky. It was a June night. If embryo No. 3 had hung on, I would've just given birth to a baby. If I had done another round of IVF, I might still be wondering if this would be the one that would work.

But after my last conversation with my fertility doctor, I'd hung up the phone knowing I lacked the motivation for more IVF. I was at my parents' that day, too, and told them I'd decided to quit. I knew my father thought I should have done so long before. "Your body is telling you something," he'd said before I signed up for my second IVF. Adoption had always been the next step in my mind. Before the miscarriages, I'd see mothers on the street with children adopted from China or Vietnam or Guatemala, and I'd feel intrigued, even a little envious. For various reasons—my travels in Asia; the values of my parents who sponsored Vietnamese and Cuban refugees—international adoption resonated with me in many ways. But after three years immersed in the world of blood and genetics, adoption now felt distant and abstract.

"I feel like I was adopted by Fred," my dad told me the afternoon I decided to quit IVF. Fred was one of my many step-grandfathers. My father's mother—beautiful, charismatic, with a wandering eye for handsome men—had married seven times by her seventies. Many of those marriages didn't last long, and Fred Levy, my father's first stepfather, was around for only a handful of years. But he was the one who played tennis with my father and asked him to sit between him and my grandmother to listen to Jack Benny on the radio. He drove from New York to Connecticut, where my father was in boarding school, to watch his football games. Until Fred died, he remained in my father's life. "Fred is in my blood," my dad told me that day. He may have said those words dozens of times before, but it was the first time I remember hearing them.

Now my husband was calling with an adoption social worker on his other line. We'd recently signed up with an agency and expected to be in for a long wait. But now this: A baby girl. Born to a Japanese student traveling in the United States. The social worker had few other details. Were we interested? My husband and I hadn't had a moment to talk about it. "Yes," we said at the same time. It was the first time in years that anything to do with parenthood seemed so obvious.

The baby had to be discharged from the hospital the next day and my husband, an attorney, had just begun the biggest trial of his career, which would last for two months. I was 2,100 miles away from the Los

Angeles hospital where our baby was waiting in the nursery. Within an hour, my mother, my father, my brother, my sister-in-law, my niece, and my nephew filled the kitchen in my parents' house. The phone rang. It was the social worker. Then it was my husband. "Who can loan us a car seat?" he asked. "What else do we need?" The social worker called twice more. In between, my husband called again to map logistics. "Change your plane ticket," my mother reminded me. The phone kept ringing; every conversation interrupted by another call from another family member who'd heard the news: congratulations or tears or questions or all three followed. My sister called with Japanese nicknames for my daughter. My husband phoned again: he found a car seat; a friend would drop it off tomorrow. Another sister-in-law ran out to Target to buy diapers, onesies, changing pads, tiny socks to fill my suitcase. My mother told me to change the plane ticket again. "What would we name her?" someone asked. I had no idea. "You're getting a special baby," my father told me. When my brother, a pediatrician, came to the house, he handed me a piece of paper on which he'd written the name of the best formula. "1 oz. every 3–4 hours at first. Gradually increase volume." I folded it and tucked it in my pocket. Aside from instinct, it might be all I had to go on for the next few days. Then I asked my brother if it was possible to know if the baby, who by all accounts was healthy, had a propensity to develop autism or other neurological diseases. He smiled. "You can't, babe. Welcome to parenthood."

"Have you changed your plane ticket?" my mother asked again. It was eleven p.m. I called the airline and got the last seat on the only flight that would get me to Los Angeles in time to bring my girl home.

I had three hours to prepare for parenthood in between when I arrived in Los Angeles and when my husband picked me up for the drive to the hospital. Soon after a taxi dropped me off at my house, a friend showed up in a station wagon filled with baby washcloths, bath towels, burp cloths, nipples and bottles, a baby thermometer, a baby bathtub, baby lotion and shampoo. Another friend who had heard the news sent an e-mail: "I'm sitting at my desk weeping. I'm dying to call you but won't since you're running out the door. If you need ANY help, please let me know. . . . I'll pick up diapers, food, Valium for you!"

My to-do list for the afternoon went like this: Buy baby Tylenol/formula/baby laundry detergent. Call editor. Ck email. Calls for article on sex trafficking. Give away tickets for baseball game. Grocery store. Call adoption lawyer. Find a pediatrician. Look up baby names.

"Are you ready?" my husband said from his cell phone on his way home. My lunch had consisted of a brownie and a Diet Pepsi on the airplane. I had weeks of unfinished work that I had to somehow put on hold. I had never read a parenting book. I didn't know anything about infant CPR or first aid. I wasn't sure if I knew how to give a baby a bottle. I definitely didn't know how to give a newborn a bath.

When we arrived at the hospital, there was no one from the adoption agency to greet us. We walked into the nursery, the borrowed car seat in my husband's arms. The room seemed cavernous, with empty bassinets lined up along the sides. A nurse worked at her computer in one corner. And in the center of the room, right in front of us, was a single bassinet with a baby inside. She was the only one in the entire nursery. Our baby. All these years it felt like there had been gatekeepers between a baby and me: doctors, nurses, my own body, my indecision. Now it was just us. Our daughter was lying there as if awaiting us, swaddled and sucking on a green pacifier that seemed as big as her round, beautiful face.

In the adoption world, they call babies like ours "surprise babies." They are the infants whose biological mother didn't make advance plans; sometimes the adoptive parents have a couple of days to pick up the child; sometimes a few hours. Someone once asked me if my husband and I could have decided that we didn't want our surprise baby after all. I'm sure we could have. But it wasn't a question that ever crossed my mind. I'd heard of international adoption agencies that allow parents to visit orphanages and pick the child they want. Who do you choose? The pretty one? The one who looks at you first? The girl? The one who is sickly and needs you the most? And how do you know how any of those things translate into a deeper or happier love? Our daughter was healthy, good-natured, and adorable—our

"Zen baby," we called her in those early months—and for that I felt extremely lucky. But in truth, the moment I heard about my baby, she was mine. I was beyond choosing her.

For a while, I wondered if all the fertility procedures and failures had changed me for the worse. I wondered if I had become less courageous, more paralyzed by anxiety of all that can go wrong. Some studies suggest that parents who have undergone IVF tend to be overprotective. Perhaps it's because they are clinging to their hard-won babies, or that the vigilance and control and fragile pregnancies of fertility medicine inevitably heighten one's sense of danger.

During the first few months of my daughter's life, I realized I had changed, but not in the way that I feared. My daughter came so unexpectedly that I didn't have time to research and worry about parenting skills and baby hazards. I avoided the Mommy & Me classes that friends invited me to join. I had limited appetite for debates about immunizations, Ferberizing, the family bed. Intellectually, these topics were interesting. Emotionally, I didn't want to squander what I'd regained through my daughter's arrival.

In part, I once again felt lucky in my life. But there was more. I finally had the tangible evidence of the gift of *not* knowing, not researching, not obsessing over a decision until there's nothing left but a whirlwind of confusion. Adopting my girl had been the opposite of all the previous years of careful, perfectly monitored fertility treatments. Her entry into our lives was pure, spontaneous romance.

Now, when I read adoption listservs or talk to other adoptive parents, I often hear phrases like "meant to be" and "fate" and "God's plan." For me, the thread that connects their stories, and their lives with their children's, isn't fate but the instinctual pull of parenthood. Not of biological parenthood or adoptive parenthood—just parenthood. In those first months, my desire for my daughter felt so consuming that it seemed hormonal. I was bleary-eyed and overwhelmed and yet I would wake up in the middle of the night and go to her bassinet just to put my hand against her cheek and watch her sleeping face.

The other evening, as my husband and I were drifting off to sleep, after another exhausting day chasing our girl, now a toddler, he murmured: "She's in my bones." Two years earlier, similar words from my father helped nudge me toward adoption. I only needed the encouragement once, though. My husband and I are making plans for baby No. 2. We're not yet sure where that baby will come from, or when. Maybe we'll receive a photo and a short medical report in advance, or nothing much at all. I will cross my fingers for a healthy, wise child, like my firstborn. And then I'll close my eyes and jump. It's the only way I know to become a mother.

KATHRYN HARRISON

Cradle to Grave

"I'M LOST," MY GRANDMOTHER TELLS ME. She's calling me at work, where, in New York, it's about four in the afternoon. She's in Los Angeles, three hours behind and three thousand miles away. There isn't any door to what passes for my office in the editorial department of a publishing company, but I turn my back to its entrance, a gap in the metal partition.

"What do you mean, 'lost'?"

"I mean I haven't the faintest idea where I am!" Behind her indignant voice I can hear traffic, freeway traffic, a sound familiar to anyone who's ever lived in L.A.—a noise of rushing air, of atmosphere displaced by velocity—so familiar that I never noticed it until long after I'd moved away and then returned as a visitor.

"Where were you going?" I ask my grandmother.

"To the dentist."

"In Pasadena?"

"Yes. Yes. You know, Dr. Bendel."

"So you're calling from a pay phone?"

"What do you think!"

"You don't have to snap at me. You were going southbound on 101?"

"Yes."

"Then what?"

"I don't know. I don't know. I missed the turn or—I don't know! All of a sudden I didn't recognize any of the exits." The panic in her voice ignites my own, that and the fact that it was just the previous weekend when I had yet another argument with my husband, who demanded to know why I still hadn't written—why, in effect, I refused to write—to the Department of Motor Vehicles to ask that my grandmother's driver's license be revoked. Was it, he wanted to know, because I didn't care if she ran over someone?

"It isn't only herself she's jeopardizing," he said.

"You just don't get it. You don't know L.A. There's no public transportation, not within walking distance. And she's alone. Without a car she can't even get to the market."

"She can take cabs."

I didn't answer him. Of course she shouldn't be driving, not at eighty-nine, teetering on the thick cushion she uses to achieve enough altitude to see over the steering wheel of her outsize Lincoln. She's nervous ("overstrung," "prostrated," or "inclined to hysterics," in her own parlance—a "family failing," she adds to a confession presented as proof of pedigree); she's lost 40 percent of her hearing in the left ear, 60 percent in the right; she sees other drivers through the smeary haze of "spectacles" whose lenses she never remembers to clean, lenses she literally butters at breakfast, putting down her toast to push her glasses up from where they've slid down her nose. Though she won't admit it, even with her hearing aids turned up she can't understand me if I don't raise my voice enough that certain of my coworkers have become an enthusiastic audience for her calls, which I receive regularly. And her mishaps are not only, or even usually, navigational. She has four aged cats, two with pancreatitis, the symptoms of which she reports in gruesome detail, along with her valiant and creative failures to disguise their little enzyme pills in lumps of liverwurst or folded squares of lox. A third has an inoperable brain tumor and is taken with "fits." The Mexican cleaning lady kindly brings her homemade tamales, which give her indigestion, which she calls to suggest is cardiac arrest. Her legs tingle, presaging cerebral hemorrhage. She has a "poisoned" toe. What does it mean, do I think, if she sees falling "twinkles" in her left eye? There are prowlers in the back and vandals in the front. All these reports interrupt and disturb me, sometimes to the point of tears. But it's the roadside emergencies that dismantle the defensive stance I maintain with regard to my grandmother—namely, that she's managing by herself—they, along with my husband's warning that the responsibility for the deaths of her innocent victims will be mine.

"What exit did you take?" I ask her, watching the LED readout above my phone's keypad, the one that counts off the seconds, minutes, even—potentially—the hours of a conversation.

"I don't know."

"Well, can you tell me what street you're on now?"

"I'm at a Standard Oil station."

"Did you ask the attendant for directions?"

"He's a lout."

"I don't care if he's a lout. Did you ask him for help?"

"A foreign lout! I couldn't understand a word he said!"

"Okay," I say, "okay." The one critical thing to bear in mind during a call like this is not to make her cry. When she does, she gets even more rattled and confused, and uses this against me. Reflexively—it's not strategy but habit—she'll amplify and prolong the noise of her crying to make me feel like a monster, to remind me, again, that if she's alone and helpless, it's my fault. I'm the one who's left her in that state. (Or, to be accurate, among those who've left her, I am the only living and therefore, practically speaking, accountable person.) "Did you try calling triple A?" I ask.

"They don't come unless the car breaks down."

"You called them?"

She says nothing, meaning: no, she hasn't, and she won't.

"What's the point of being a member? You do pay annual dues, you know."

Silence, except for the ambient noise of cars rushing past. Could the complaints of conventional parenthood be very serious, really, in comparison to having become the de facto guardian of an irrational and contrary octogenarian? I wonder if, from my desk in New York, I could call the Automobile Club of Southern California and convince a dispatcher to send someone in a tow truck to find my grandmother and drag her home. A long shot, at best.

"Okay, Nana," I say. "This isn't an emergency. Find out what exit you took. And I need to know which freeway. Read me a sign. Ask the lout if you have to. Or get hold of a different lout, one who speaks English."

It takes time to guide my grandmother back to her house, nearly an hour when I add in two more calls from pay phones en route. But I do get her there. I've driven her to the dentist so many times that I know the tangle of freeway interchanges between her house and his office well enough to guess which wrong turn she took. Having grown up with my mother's parents, lived with them in Los Angeles until I went to college, I can navigate the streets of that city, even from a distance. But I can't stop my grandmother from losing her way, or from talking to strangers. Or, and especially, from hitting another car.

We've been married three months, barely, when my husband makes an announcement. My grandmother, he says, is about to develop from a problem into a crisis. He's not being an alarmist: he sees it coming. What's more, he knows that the best way of averting what will otherwise prove inevitable disaster is to move her East, to live near—with—us.

"You can't be serious," I say. He's offered this outrageous suggestion in the course of one of our unhurried, postprandial walks through our neighborhood, block after block of graceful old homes, brownstones, mostly, some with bay windows before which we pause to picture ourselves on the other side of the glass. I pull up short. He stops, too, and looks at me.

"Why not?" he says.

"Why not!"

"Yes. Why not?"

"You have no idea what you're saying. You've never lived with her. You don't know what she's like."

He waves a hand through the air, as if dismissing a gnat. "She's an old woman."

"What is that supposed to mean?"

"How hard could it be? I mean, really?"

"Oh my God! Why do you think my mother fled? What do you think I'm doing on the opposite coast?"

Just because my grandmother is now ninety years old, less than five feet tall, and eighty-something pounds, is not reason to underestimate her power to drive a person crazy. Demanding, conflicted, inconsistent (but not consistently so), histrionic, given to gestures of extravagant (and often but not always self-interested) generosity, insecure, witty but not wise, manipulative, suggestible, and, yes, very charming and funny, she's the life of the party, as long as it's her party. "It's like this," I try to explain. "You can't even sit on a couch with her. No matter what she's doing, no matter how apparently quiet and still she's being, you can't be in the same room with her and manage to read a book. Or even one page of a book. Or think a single coherent thought. You haven't spent enough time with her to find this out, but she's the human equivalent of a black hole. She just wants wants wants and you can't ever give her enough. She pulls you in and consumes you. Even when she isn't making a sound she's loud. Deafening."

"Don't you think you're being a little dramatic?"

"Only to make a point. She's not a grandmother the way people like to imagine grandmothers. She's not the Waltons. She's more Auntie Mame meets the Addams family." I picture her standing in her kitchen, tirelessly, laboriously, cutting up raw beef heart to feed to her cats. It takes so long that she's always interrupted at the task, and she forgets to wash her hands before answering the phone or responding to the shatteringly loud buzzer on the clothes dryer, a buzzer you can hear from the street. The trim around the doors, painted white, the wall where the phone hangs, also white, the doorknobs and cupboards and the cutlery drawer—the laundry itself, freshly washed and dried—everything is anointed with bits of raw flesh that dry and stick with a fierceness rivaling that of epoxy. From experience, I know that to pick even one surface clean requires dedication—were the cleaning lady to attempt removing all the meat, she'd have time for nothing else—and so the house always looks as if it's recently hosted a murder, a messy killing at the hands of a psychopath unconcerned with being found out.

She might be amoral, my grandmother. At the least, her morals don't align with the rest of the world's. "Listen," she said to me when I was twelve. She'd backed me into a corner of the linen closet. "If you get into trouble, I don't want you to have an abortion. Give me the baby." I said nothing. "Do you understand?" I nodded. "Promise?" I kept nodding. Not only was I years away from losing my virginity, I hadn't even started menstruating. When I did, and—still a virgin—made the move from sanitary pads to tampons, she slapped me across the face and called me a tart, having apparently interpreted my choice of feminine hygiene products as one that had necessarily followed on previous penetrations: a tacit admission of promiscuity. Is it any wonder that as I grew up I never played house or dressed dolls or saw myself presiding over a family? Taken together (and there was no vantage from which I could see them as individuals), my grandmother and mother presented more of a cautionary tale than a role model.

Pregnant at seventeen, hurriedly married and, as soon as she was delivered of me, pressured into divorce (my father, having provided the one thing my grandmother might, grudgingly, have accepted from him—stud—disappeared), my mother left home but never managed to put more than a few, fraught miles between herself and my grandmother. And since neither of us—not my mother, nor me—invited friends home to see the meat on the walls, or hear my grandmother's screams of rage when we crossed her, no one ever understood what we meant when we said she was "difficult." If anyone met her, it was at a dinner party or a restaurant, a public venue where, on the arm of my tall, distinguished, mercifully (defensively?) deaf, and introverted grandfather, she told antic, exotic, and often very funny or dramatic tales of growing up in China, taking a refugee Russian prince as a lover in Nice, and other adventures involving her eccentric relatives, of whom a disproportionate number were sexually deviant and/or inclined to wave firearms and threaten suicide over minor romantic mishaps. There was, too, a de rigueur escape from Nazis, complete with slavering "Alsatians," which sounded more foreign and therefore more thrilling than "German shepherds." Undeniably, she's bewitching, right down to her poisoned toe.

"Look," my husband says. "If she doesn't run over someone, she's going to fall down the stairs and break her hip. Or go off the deep end and leave all her money to her cats."

"Fine. Let her."

"When she breaks her hip or wrecks the car and gets taken to court, or when her heart finally gives out, it's going to be a lot worse to deal with from a distance. Think of what it was like when your mother was dying, or your grandfather. Weren't things difficult enough when you were there, in the same city?"

We keep walking, both of us silent. It's January, dark with ice indistinguishable from the stone sidewalk underfoot, to the eye anyway. The cold comes through the leather soles of my shoes and makes my feet ache. I know what's coming next, it's only a matter of time—two blocks, as it turns out—before my husband reminds me of our last visit to my grandmother's house, when he found shingles blowing off the roof and an open box of blank personal checks left on the back porch, its contents carried off and scattered over the ill-tended property by the same wind that was prying up the shingles. It's hard to guess which form of disability my husband might find more disturbing, my grandmother's failure, literally, to keep a roof over her head, or her financial irresponsibility.

"Here's another way to look at it," he says as we turn the corner onto our own block. "If we were to help your grandmother put her house on the market, sell it, and move her East to live with us, then we could pool our resources. Together, the three of us would be able to put enough money down on a house to carry the balance for about as much as we're spending on rent."

"Aren't you forgetting something?" I say as my husband unlocks the door to our apartment.

"What?"

"She doesn't want to live with us."

"Not even if she had a separate floor? Most of these houses are carved up into two or more units, anyway."

"She hates New York."

"She hasn't been here in, what, thirty years? Forty?"

"Okay, she thinks she hates New York. It's the same difference."

"She could bring her cats."

"Even so."

Ironic, given my apprehension about living with my grandmother and her insistence that she'd rather die alone, in agony, than move to the East Coast, that my standard response to her panic calls becomes to suggest that none of it—not the dead car battery, not the malingering pet, not the overflowing toilet or the clogged rain gutters or the carrot peeler stuck in the garbage disposal—would present much of a problem were she living under the same roof with me and my husband. But comfort and safety are not, finally, the enticement she requires, and we both know this; she does and, on some level, so do I. There is an offer she can't refuse, but I haven't made it, not yet.

"I don't know a single soul in New York!" she says. "Not a soul!"

"Nana. You don't know anyone in L.A. anymore, either."

"There's Eula."

"She's in Oxnard. Sixty miles away."

"Crystal."

"You just told me she's dying of stomach cancer."

"And Rosa. What about Rosa?"

"The cleaning lady doesn't count. And neither does the doctor or the dentist. Or the gardener, who, by the way, is ripping you off. He's stolen every tool out of Poppa's shed."

"So why don't you move back, if you're so worried about me?"

"Because I'm twenty-eight. I have a job. Here. In New York. We both do. You're ninety. You're the one with the house you can't take care of, and the sick cats you can't get to the vet, and the car you can't drive without getting lost. Why can't you be flexible for once in your life?"

My grandmother slams the phone down. I look up and see the marketing director's assistant leaning in the doorway. "Your nana?" she asks. The editorial department has grown bored with my calamitous grandmother, but she's just catching on in other divisions.

I don't have to ask the DMV to take her license away. It expires, and she fails the written test required for its renewal. But it's not the practical obstacle of my ninety-year-old grandmother not being able to drive herself around that threatens me. My husband is right—she can take cabs. But to allow her to be frail, to depend on me in her old age as I, during childhood, depended on her, in this I'm forced to recognize the loss of her license for what it is, a harbinger of her death. Despite her careless cruelties, despite her having subjected my teenaged mother to psychological strafing, and then, after taking me on, presided over so bizarre a household that she blighted whatever chance I might have had at being a socially tolerated teenager, she is my family, my only remaining family. I know—I haven't forgotten—that many times, every day of my early life, it was she who stood between me and the world, she who was the parent my mother couldn't be. If she failed us, it was in the force of her desire, her need. She wanted my mother and me, wanted us absolutely and irrevocably, for herself, forever, and she didn't protect us from that hunger.

She's still ravenous, insatiable. And it's life she wants to consume—new life. There isn't a pram or stroller she doesn't chase for a chance to ogle the baby within, not one young mother she doesn't pump for details of her offspring's birth, its sleeping habits and digestion, the new tooth breaking through its wet smile. Every summer of my childhood, we visited the San Diego Zoo. While my grandfather and I went off together, riding the tram from one exhibit to the next, my grandmother stood alone at the big viewing window of the primate nursery. Rapt, she watched the keepers diaper and feed infant chimpanzees, orangutans with their halos of orange fuzz, and, once, a little lowland gorilla with its strangely exhausted face, black bags under its eyes. She gasped audibly with delight when one of the blue-uniformed young women put down a bottle of formula to lift an ape over her shoulder and burp it. "Do you want one?" I'd ask my grandmother, feeling almost jealous—her fascination was that intense—and she'd hug herself, ecstatic. "Bless their little furry hearts!" she'd say.

"I know how to convince her," I tell my husband. It's a weekend morning. We've just come home from the newsstand with the *New York Times*. He looks up from the paper and raises his eyebrows.

"I'll tell her that if she comes East, we'll have a baby. I'll stop using birth control as soon as she moves. Otherwise, who knows how long it might be before we could afford a bigger place? She might not live to see a great-grandchild."

My husband looks at me, nodding slowly, frowning. "Do you think that will work?"

"I know it will."

My husband stares at the wall, the newspaper, still intact, in his lap. "Would we, do you think? Would we have a baby, just like that?"

"I don't know. I mean, how would I know?"

We've never discussed having children, not really. I think we assume they'll happen as marriage did: without deliberation. It's not that either of us is careless or casual—far from it—but three days after our first date, my not-yet husband, with whom I'd spent an aggregate of six or seven hours, handed me the key to his front door. And, without hesitating—without thinking—I took it; I moved in with him. Over the years, we've discovered that with us it's always like this. No matter how large the transition, our approach to it feels as if it has been choreographed, steps we learned so long before, we can't remember when we didn't know them.

"You'd go off birth control?"

"Uh-huh."

"And if you went off birth control . . ." He stops speaking, his eyes not focused on anything in the room.

"Six to twelve months."

"Six to twelve?"

"They say that's the average. The average time it takes. Or the time it takes the average couple."

"To have a baby?"

"To conceive a baby."

"Huh," he says. He pulls out the Metro section.

Were anyone watching us, standing as we ourselves have stood outside a stranger's window looking in, he or she might imagine us to be discussing an article from the paper in my husband's hand, for all the anxiety we betray.

"Well?"

"No."

"No?"

"No."

"Come on, Nana."

"I've lived in southern California for fifty years. Of all the places in the world I might have settled, I chose here."

"I know. But—"

"Fifty!"

She doesn't let me talk but flogs us through the familiar story of her visit, at seventeen, to Pasadena, where her father had taken her—by ship, from China—for a change of scene, i.e., separation from a problematic suitor. He bought her a chestnut quarter horse, had it outfitted with a Western-style saddle that was trimmed, like the bridle and reins, with Mexican silver, and watched her tear through the orange groves on the animal's back. Somehow he finagled an invitation for her to ride in the Rose Parade on New Year's Day, 1916, the pretty mare's mane braided with ribbons, her neck wreathed in flowers. Even in the context of an argument, it's a story I love, and not only for my grandmother's evocation of a primeval, unspoiled California. Her affection, still so palpable, for her father—a young man from Baghdad who eschewed family connections to make himself into a fabulously wealthy banker by virtue of his own wits and talents—changes her voice, softens it—her. When she talks about her father, she becomes, I realize as I listen, less hungry-sounding.

It's not the romance of young parenthood I'll imagine sharing with our children when they're grown, but my interest in having a baby is quickened by my determination to have my way over my grandmother. And, though my first child will be a teenager before I understand this, for me the idea of pregnancy answers a narrative impulse: it promises a way to recast the circumstances of my own birth. As did my mother, I will tempt my grandmother with a baby, but my child isn't to be a surrogate, a stand-in for myself. I plan to hold her—him?—tight and thus compel my grandmother to my side, bend her will by force of blood kinship.

Pride insists my grandmother not concede immediately that the birth of a great-grandchild might trump any other pleasure left for her. In truth, neither of us wants to face the symmetry of our relationship: just as it was she who protected my infancy, who bathed and fed and diapered me, it is I who will guard her through senescence and be midwife to her death.

I am all she has, and this frightens both of us. So we argue geography.

"Fine. You like California—"

"Love."

"You love California. But it's not as if you go to the beach. Or do business with movie producers. You don't grow strawberries or manage a motel on the palisades. And you don't drive."

Her voice acquires a stronger, almost plummy, British accent. "I suppose it's foolish of me to imagine that you might understand what climate means to someone who was raised in Shanghai."

"You hardly ever go outside!"

"I look out the window."

"There are windows in New York."

"Yes, but not any I want to look out of."

"How do you know?"

She coughs in lieu of answering.

"Anyway," I say, "all this is beside the point. The bottom line is, if you come live with us, I'll have a baby."

"And you won't otherwise? For you, having a baby depends on where I choose to live?"

"Yes."

"Don't be absurd."

"As far as you're concerned, it does. Because no matter when I have a baby, if you're in L.A., you'll hardly ever see her. Or him."

Silence.

"Well?"

"I'm not discussing it any further."

"Fine. We don't have to discuss it. Are you going to move?"

"No."

"Nana."

"No."

"Are you?"

"I'll think about it."

"I knew you'd say yes!"

"I didn't say yes. I detest New York, and I didn't say yes."

I'm so practiced at fighting with my grandmother that I don't have to scheme—it's second nature. And the thing to do now is nothing. After a day or two of silence from my end, she'll call to say her cats' claws need cutting, or the oven doesn't go on anymore, or she's lost a tax document she needs. Under one or another pretext, she'll wait for me to bring up the idea of the move, which I won't, and my not mentioning it will make her nervous. It will inspire her fear that perhaps I've reconsidered; perhaps I don't want to take her on, after all. Mean, admittedly, and efficient. I know what I'm doing.

But when she calls, it's without an excuse, and after a few minutes of small talk—in her prime, she'd have lasted far longer—she says, "I've given it some thought."

"Given what some thought?"

"Your suggestion."

"Oh?"

"I've decided to do it."

In apology for having forced her surrender, I don't gloat. "Really?" I say. "That's great!"

"Yes."

She's very cool, very casual, pretending, perhaps, that it's just a visit, an extended visit.

Which, as it turns out, it is. In eighteen months, my grandmother's rheumatic heart, its valves damaged by a childhood bout of scarlet fever, will at last give out, so enlarged at the very end that I'll be able to see it beating beneath her ribs from across the room—not her pulse but the muscle itself, as it clenches. Still, she will have become a great-grandmother. As obedient as when I was a child in her care, my very flesh unwilling to disappoint her, I keep my end of the bargain and get pregnant almost immediately, eliciting no more surprise from the father-to-be than "I thought you said twelve months."

"Six to twelve. On average. I must be the woman who balances out someone at the other end of the curve, one of the women who take a year or two."

"I guess so."

"I know it's fast," I say into his shirt, hugging him.

"All the more reason to get a babysitter sooner rather than later," he says.

We should have hired a sitter for my grandmother at the outset, but we wait until it's obvious, even to me, the one who doesn't want to see it, that my grandmother can't be left alone in the house. In the ground-floor apartment of the brownstone she helps us buy, she has a soft, new couch (a loveseat, technically, but it looks like a couch when she sits on it) and a big television that she won't stay put in front of while we're at work. Instead, she roots compulsively through boxes of old papers, rereading and tearing up letters—"Saving you the trouble," she says tartly when I ask why—and going back and forth to her tiny kitchen to make tea, forgetting the tea, and burning the bottoms out of five or more kettles. She trips over the corners of rugs or the tails of her cats, all four of which fly East (accompanied by pet courier, for an air fare higher than her own). She spills kitty litter over the hardwood floor, inviting a fall.

There are enough little emergencies that she calls me at work several times a day to announce what it is she's dropped or broken, or where she's fallen, or that she can't get the faucet to turn off, or on, or that she's lost a cat, or broken a tooth: the list is long, then longer. I, in turn, call my husband, who (to spare me, in my morning sickness, another subway commute) forfeits his lunch hour to go home and sweep up litter or discard the most recently burned kettle or, once, lift up my grandmother herself after she'd fallen, mid-conversation, and dropped the receiver where it picked up her cursing me and my chivalrous husband as she crawled from one corner of the little apartment to another, looking for something with which to pull herself up from the floor. "Left me. Left me all alone," she spit, just as savagely as she used to fifteen years before. "Dreadful girl. Chit of a girl. How could she. Leave me. To marry that lout."

It's predictable, so why don't I predict it? When the baby comes, a little girl, she isn't happy. She's jealous. For by now she is the baby, my baby, the one whom I summoned to my side so I could watch over her, so I could make her breakfast and dinner and bathe her each morning. And diaper her—that, too, toward the end. Not that she admits to envy. When my grandmother was four, she had a baby brother whom she wished ill, and after that he died. In her heart, if not her head, she believes that jealousy of babies is wicked, and will be punished. To banish the feeling, she criticizes me. With respect to my little girl, I do everything wrong. I give her the wrong name. I spoil her by going to her when she cries. I tease her by encouraging her to reach for a toy. I use the wrong baby soap: that's why she has a rash on her cheek. Worst of all—both for the intimacy it betrays and for the insult of my giving away what she believes to be her own—I nurse my baby. Rather than feed her formula from a bottle, I give her myself, my body. I allow (encourage!) this interloper to eat me up.

"It's plebeian," my grandmother says of breast-feeding. It's what animals do. Aboriginals. Destitute mothers who can't buy formula. "We saw them in the street in Shanghai, you know. Women with their breasts pulled out until they looked like razor strops. Their filthy children ran alongside them without letting go." She shudders fastidiously. "No woman who cared about her décolleté would consider it."

But I don't stop. Nursing is one of the few things I can do while sitting with my grandmother on the couch, both she and the baby inclined to doze, freeing me, at least mentally. With the baby suckling in my arms, I listen to my grandmother breathe as she sleeps, air whistling faintly as it moves through her chest. The room fills with the small noises of bodies, hers as well as the baby's. One of my grandmother's hands rests on the couch between us, and I take it in mine, surprised as always by the heat of it, as if she had a fever. On her forearm are two new scars, pink and square. My fault—trying to spare her pain, I pulled off a Band-Aid too quickly, and her skin, so thin it's transparent, came off with the adhesive. She didn't cry, not even at all the blood, but I did.

The symmetry, the one we didn't want to see, is unavoidable now. Having shaved away hours once reserved for sleep to care for my grandmother as well as my daughter without having to quit my job, I am always tired, and easily brought to tears. I get up at five forty-five to nurse my daughter—nine months old—until six-thirty, when I take her downstairs with me to rouse my grandmother, always hesitating at the door to her apartment, praying *Don't let her be dead,* or, sometimes, *Please let her be dead.* I find her awake in her bed, waiting to be taken to the bathroom, to be stripped and bathed and dried and diapered, then dressed in a fresh housecoat and led to the couch where she waits for her breakfast. As it is, my husband goes to my grandmother during the night—we've set up a baby monitor in her room so as not to miss her cries—so I try not to wake him before eight, when he has to get up for work. The baby crawls around the apartment unsupervised while I tend to my grandmother in the bath. This is okay, I think: there are no stairs to fall down, the electric outlets are baby-proofed, the cabinet under the kitchen sink emptied of cleansers. One morning, however, when I emerge from the bathroom with my grandmother, I find my little girl sitting next to the cats' bowl, her pink cheeks crammed with Friskies.

"You see?" my grandmother cries, delighted. "You see how hungry she is? It's rubbish, this idée fixe you have about formula being unnatural!" The baby cries when I try to empty her mouth of the cat food, pleasing her great-grandmother. She grabs at my wrist with her little hands and turns her face away.

I must have known all along that it might come to this. That I might have to choose between them. My husband's parents arrive for a visit and cannot hide that they are shocked by the state of the house and its occupants. We think we're managing, they tell us gently, but we can't continue as we have been. The worry on their faces shows us what a mirror can't. It isn't just the circles under his eyes or the cold I've had for weeks. All new fathers are tired, but not so tired that when they get down on the floor to play with their babies they fall asleep. And a lot of women lose weight while breast-feeding, but not like this. We've engaged a practical nurse to help care for my grandmother in the mornings, and a full-time babysitter does whatever housework she can during those minutes not taken by the baby, but the end of each weekday still confronts us with dinner, laundry, dishes, and two people whose needs are overwhelming. And weekends are worse; it seems impossible that two days can last so long. Or that, living in the same house and sleeping in the same bed, my husband and I see so little of each other.

"I can't," I say to my mother- and father-in-law. "How can I, when I promised her? I promised, no matter what, I'd never put her in a home."

"But is this what she'd want for her great-granddaughter?"

No. Not her old self, anyway. The woman who took care of the child I was would think it wrong for a baby to be passed back and forth between parents often too tired to smile. But the frail and failing creature she's become might not agree. To whom do I owe the best of myself? My grandmother, who, sixty-two when I was born, remade her life for me? Or my child, who has been in the world—outside of my body—for little more than a year? Fourteen months, that's all, since my heart was hung above her head, filling the dark with its sound.

The words the social worker uses to describe the place—"convalescent hospital"—supply us with the lie that makes it possible for me to betray her—for her to accept what she recognizes as my necessary betrayal. After being hospitalized for an ailment that fails to end her life, my grandmother is transferred to the Greenpark Care Center, where she goes to "convalesce."

The first few times I visit, I come alone. Then, one Saturday, I bring my daughter. Eighteen months old, no longer a baby, she runs past wheelchairs in the corridors, eluding skeletal hands that reach out to touch her arms, to feel their plump life. When we enter my grandmother's room, she stops short of the bed. The two of them regard each other solemnly. Why would my daughter's great-grandmother be any less an object of fear, or revulsion, than the withered, toothless crones that snatch at her in the hall?

"Growing," my grandmother observes with a curt nod, and she asks where I bought the dress my little girl is wearing. Then she changes the subject, less inquisitive about her only descendant than I remember her being about those anonymous babies in buggies pushed by strangers. I wait a few weeks before trying again, but once more my daughter's presence makes the two of us awkward with each other. In this small room smelling of disinfectant, neither of us can avoid consciousness of the mortal transaction my firstborn represents. Was it—was I?—cold-blooded? Was I kind? Is it possible for me to have been both? I don't bring my daughter back.

Along with maintaining the fantasy that it's only a matter of time before my grandmother returns home, when we're alone together, she and I pretend that home doesn't include my daughter. Of course, we never say she doesn't exist; we just don't mention her. Instead, our conversations take us back, past my childhood and past my mother's, back to my grandmother's own beginnings, a world and a lifetime apart.

"Isn't it remarkable," she says, "how much I remember?" The names of all the teachers at the little day school in Shanghai's International Concession, the name of the seamstress who came to the house to measure her and her sister for dresses. The white rabbits she kept as pets, their names, and the names of all their offspring. The brand of talcum powder preferred by her governess, and the day of her birth. The Presbyterian minister who lived down the street, his wife, their four daughters, and all their names and birthdays. In fact, she is remarkable with birthdays; she remembers every one she ever knew.

Except mine. On the day I turn thirty, the month and day writ very large on the wall calendar in front of her bed—one of those aggressively outsize calendars hospitals use to preserve patients' relationship to the passage of time—my grandmother takes my hand and pulls me close and asks me, did I forget to buy a card for a cousin whose birthday falls three days later. "No, Nana," I tell her. "I did get one. Don't you remember? You signed it." To hide my face from her, I step into the hall.

It doesn't matter, I tell myself. Don't let it matter. It's a part of the bargain we struck, that's all. I'm not sure why I conclude that this is so, but I do. The thought arrives with the weight of truth, and I dry my eyes. How has my grandmother's failure to wish me a happy birthday been transformed from injury into something more like a minor side effect, painful, perhaps, but without importance?

Back inside her room, sitting by the bed and stroking her arm, I am trying to understand. It isn't frustration at not being in the position to buy me a gift—for years she's had me pick out my own birthday card and fill in all but the signature blank of the check she enclosed—but a refusal to acknowledge what day it is. Somewhere in the calculus of births and deaths, in the trading of homes and the reversal of roles, in the loss of her daughter and the arrival of mine, this meaning of March 20 had to be erased. I don't fully realize it yet; I won't allow myself to put it into words: the fact that I am no longer anyone's child. But I will. In time, I will.

LAURIE ABRAHAM

Mother's Little Helper

LEAVING THE COSSETED WHIR AND HUM of the high-rise office building where I work as an editor, pushing through the revolving doors into the cold night air, I think of it: my drink. I've got half of a bottle of white open at home, I know, and I'm pretty sure there's still a Brooklyn Ale in the refrigerator. . . . There's definitely some beer in the pantry, so I can always pop one in the freezer for a few minutes. But the delay—too bad I can't call my babysitter and ask her to do it for me, but I'd never. She might think I *need* a drink. I want to be blithe, casual about my after-work libations: Oh that little bottle of beer that somehow got into my hand, how did it get there?

On the subway, in the dirty yellow light, I keep working, nipping and tucking with my pen. Striking a word to save a line. *Yes,* perfect! Where am I? Only at Broadway and Lafayette, I've got time. I know in my body how long it takes to get from Midtown Manhattan to Brooklyn, but sometimes I get too engrossed and miss my stop: Smoothing out the rough patches, cutting, tweaking, making it fit, making it fit.

"Excuse me," I say. "Excuse me." Dragging my heavily laden purse and bag out of the train, I'm back outside. A block from my house, I pass the Cuban restaurant with an elegant wooden bar, so smooth you could lay your cheek against it. Then, a few storefronts down, the real bar. Several of its windows are actually painted over. I peer into the blue-blackness when the door swings open: Who comes to this place? What are their lives like? Are they still single? Tiny white lights twinkle and beckon, and the bar itself is illuminated by a spotlight, the star of the show.

Around the corner, and I'm home. "Mommy, Mommy!" My daughter runs into the hall. I crouch down to hold her. I'm squeezing her, asking about her day, but then I've got to get into the kitchen to bestow an equally enthusiastic greeting on my baby, who's adorable and relatively patient. Since I'll pass by the refrigerator on the way to the highchair where the baby eats dinner, I could stop and flick the cap off of a cold, cold beer, but I won't. Somehow, it would seem like there was something really wrong with me if I opened a beer before I hugged both my daughters.

Wrong with me? Is that what this is? Let me say right off that the most I drink in an evening home with my children is two glasses of wine, and if it's beer, no more than one. But I'm obviously preoccupied with getting that first drink—I'm so relieved once I've managed to procure it. When my husband and I decided to stop using birth control the first time, we knew being parents wouldn't necessarily be easy. It would limit the time we had together and the time we each had to ourselves; it would disturb work-focused identities that were thirty-five years in the making; blah, blah, blah. It's a truism that you can't really know what it's like to be a mother or a father until you've got that small being breathing in your home, until you spring out of bed at the sound of a cry and then realizing the wailing is in your head; the baby is actually peacefully resting. Only gradually do most of us begin to recognize—perhaps because we can't bear it at first—the profound and myriad ways in which our lives have been transformed by parenthood, the stinging losses *and* the viscerally splendid gains.

At my all-mother book group, where, proverbially, swilling wine is half the fun, we joke about having to have our alcohol, but if I was to break the pact of bonhomie and ask anyone if she really "needs" it, the answer probably would be, "Of course not." Yet in more private settings, many of my mother-friends have remarked, in voices a little bewildered, a little apprehensive, that they drink more since they've had children. "I feel like I'm a bad mother because all I want to do is sit in the glider with a glass of wine and watch him play on the floor . . . I wait until ten each night to have a drink; I don't want to waste it before then . . . Around five I start tasting that first Chardonnay . . . The other night, I called my husband before he left work and said you've *got* to stop somewhere, we have no wine or beer in the house. He asked me which one I wanted, and I said 'Both.'. . . Since the baby, I definitely drink more than my husband does."

I wonder if we're all actually indulging more as mothers, or we're just more aware of it. Here's a little ditty my husband and elder daughter came up with while I was pregnant with my second child: "Mommy's thirty-nine, she's pregnant, and she drinks wine," they'd sing laughingly, gaily. (I did not stop drinking altogether during my pregnancies, which troubled my husband and inspired this needling, though, I have to admit, pretty funny tune.) Now, our second girl is fifteen months old, and she'll grab for my beer—she wants everything everyone else is drinking. No baby bottle for her! I keep holding on to my bottle, but I help her put the cold glass in her mouth. She's delighted, until she bumps her new, Chiclet teeth against it. Ouch! I extricate the beer from her clinging hands. She howls in protest. Mommy's beer bottle (and morning mug of coffee) are as much a part of our lives as sippy cups and Elmo.

Remember those sappy cartoons from the 1970s, with that naked, doe-eyed boy and girl, "Love is . . . Getting Butterflies Every Time You See Him," or " . . . All You Need"? Maybe "Being a Mother is . . . Staying Bright-Eyed and Sober." There's a cultural taboo against inebriated mothers. As for inebriated fathers, that's just what fathers do—in the worst cases, the family suffers, but not too horribly if the big-hearted mother is there to sacrifice all for her brood. Have the mother join the father down at the bar, however, and the family members fly off into black, lonely space. "Drunk mothers" are so unspeakable that when I Googled the phrase, the first thing that came up was a porn site featuring porcine women performing obscene acts with beer bottles stuck in their mouths. (The fat signals motherhood, I guess.) There are self-help books that address alcoholic mothers, and attest to the unique devastation of (barely) being raised by a woman whose head is emphatically elsewhere—"My mother chose alcohol over me" is a typical child's-eye view—but again, I'm not really talking about getting blotto.

Perhaps my thoughts turn to stumbling, slurring mothers because I worry that it's a slippery slope from seeking a pleasant tingle to wanting to go completely numb; on occasion, I *do* sneak in another quick pour of wine. But I'm not overly worried that I'll return to the heavy drinking of my teens and early twenties. I've got too much that is compelling and sustaining in my life now—my family, my work—to waste whole days nursing hangovers. (I know good fortune doesn't come with a guarantee, but, thankfully, my neuroticism does not manifest in dwelling on nightmare scenarios.) At the same time as it soothes, the bottle reproaches me, but not as a drunk in the making as much as overall dubious mother-material. It seems to announce that I don't like being a mother, or at least I don't like it enough of the time. Otherwise, when I'm with my daughters, why would I need to drink?

When I'm feeling lively and content or merely want to treat myself, I pour my golden elixir into one of the small, colorful Moroccan glasses my husband and I got as a wedding present. I keep these juice-size glasses gathered together on top of the kitchen hutch; I can always see them this way. They're like Christmas lights, candies wrapped in pink and green and blue foils, friends. When I'm sullen or distracted or angry, I use the cheap stemware I got right after college, the ones you could throw at a wall, pick up, and use. If I'm in a really bad way, I'm not so careful to smoothly make the transition from mommy-happy-to-see-her-girls to drinking-mommy-happy-to-see-her-girls. I grab the bottle from the refrigerator and glunk, glunk, glunk, my medicine sloshes out.

This delightful nightly ritual, this fetishizing of tiny glasses, it's a comfort, an escape, or, more precisely, a comforting escape: I'm the woman and girl I once was, the one who had infinite possibilities before her and could decide at the last minute to while away her evening in a bar. Since early adolescence, alcohol's been my breakout, my rebellion; I was a conventional girl—I did the right thing, grades, sports, mountains of extracurriculars—but I got "totally wasted," as we used to say, on the weekends and, every now and again, before school (how did I stomach wine and lemonade at seven a.m.?). Occasionally, I got caught, which variously meant running the bleachers in track, being benched in basketball, and being suspended from school (with my mother working at my high school, living a double life wasn't easy). And I can still feel the fuck-you frisson of adolescence when I'm standing, one hand (metaphorically) on my hip, the other holding a bottle or glass—especially in the presence of my new "authority figure," my husband, a virtual teetotaler.

"A mother hates her baby [because] the baby is an interference to her private life, a challenge to her preoccupation," analyst D. W. Winnicott says in his famous list of eighteen reasons why mothers "hate" their infants (his overarching message being that the mother must be aware of her inevitable hate so as not to lash out at her small child and, for an older one, so as to try to "objectively" determine whether her hate is actually provoked by the child's actions—and should be addressed—or unrelated to him). The rebellion goes both ways, then: I'm the girl rebelling against her mother, and the mother rebelling against her girl. When I'm drinking, I'm more than my daughters' mother, I'm myself. Does a woman's "self" ever fully absorb her new role, so that there is no distinction between being her and being a mother? I wonder.

Then—did you think I would never get here?—there's the buzz. I was almost thirty-six when I had my first child, almost forty when I had my second, and I'd become accustomed to a certain level of quiet, a background stillness. With both children clamoring for me, grabbing at my clothing, with the baby in my arms rearing back and practically knocking the breath out of me while the older one ceaselessly chatters, my nerves get jangled. I'm old. Whether the effect is physiological or psychological, a glass of wine calms me: *Okay, I can handle this if I just have a drink first.* A drink is the modern (time-worn?) "mother's little helper": our generation's answer to Valium.

A drink is also a shortcut to turning off efficient, tough-minded worker, turning on pliable mother. The second I put the key in the front door, I must leave behind the staccato pace and intellectual demands of work (where, if you have any clout, you make requests and see them miraculously fulfilled) and embrace the bogginess of home, where patience, flexibility, and empathy are the virtues. As Arlie Hochschild pointed out in her excellent book *Time Bind,* for many parents these days, the office is far less emotionally taxing than home, despite all our clichés about home as a refuge. Sure, I could clear my mind by doing yoga or jogging instead of drinking, but who has the time for that stuff?

God, I hate comments like that, when I hear other women throw up their hands and groan with sitcomish resignation, "I don't have a minute for myself." Too often, they seem weirdly proud of their long-suffering state, almost smug about it, and I always think, *If you want time, you'll make it.* But I don't do it, I don't make time. I'm probably as annoyed with myself as anyone.

"I'd never met a [French] mother—working or not—who didn't have time to read a book, have lunch with a friend, or go to dinner once in a while," Judith Warner writes in *Perfect Madness: Motherhood in the Age of Anxiety,* her best-selling book comparing motherhood in America and France, where her eldest daughter was born. It's not only the vast array of excellent social services the French government provides that eases mothers' burdens, Warner argues, but the cultural conviction that what makes for felicitous family life is the adult woman's happiness. With a relatively content grown-up at the center, good things will flow. Back in the States, Warner runs herself ragged carting her daughters to (soccer, ballet) classes, outfitting her home like an arts studio, trying to find decent schools and affordable child care and a few hours to work. Then, she writes in a sudden one-sentence paragraph, "I started to drink Calvados in the evenings."

In other words, the way we construct motherhood here—with the expectation that women will both parent intensively and work intensively (the affluent woman, for selfish fulfillment; the poor one, to pay the bills and set an example of industry)—*not* becoming frazzled and guilty takes great reservoirs of creativity and strength of will. And most of us can't get off the merry-go-round, at least not for very long. So, we drink.

Finally, there is pleasure (which the French don't seem to have as much problem with, either). Drinking gives me pleasure. Simple as that—or I wish it were as simple as that. In the abstract, I know there's nothing wrong with needing pleasure, and getting a drop of it from wine. And yet: *You drink almost every night; the alcohol calls to you when the clock strikes six, you* need *it.* I do need it. In our culture, to need is to be dependent, noxiously needy. Okay, so maybe I only *want* pleasure, but that sounds suspect, too. To want pleasure—it's unmaternal somehow. The domestic woman receives—gracefully and gratefully. So, again: *What's wrong with me? Why aren't my children, my nice, tidy foursome, enough?*

My children bring me pleasure, startlingly, when I least expect it. The other day, I took my almost-five-year-old out for ice cream after school. We sat on high stools, facing a bustling commercial street, sharing our Dutch chocolates with rainbow sprinkles. As we sat, she sang in her hoarse, childish voice a new song she'd learned, told me very earnestly about her love for a classmate—whom she one day planned to marry "if I don't find another boy"—and subtracted nine from twenty-five ("in my head!") to announce that there were sixteen days left until her birthday. And then she put her mouth down over the chocolate peak, her whole mouth, enveloping, capturing the milky sweetness. It flickered through my brain that I wanted to do the same to her, to put my whole mouth over her, to capture her sweetness. "[The baby] excites [the mother] but frustrates—she mustn't eat [her] . . ." is another of Winnicott's eighteen reasons why mothers hate their infants. I'd say it's also one of the reasons mothers love their infants beyond all measure. The desire a child invokes—desire, by definition, is never entirely satisfied—has to be one of the deepest, most stirring pleasures known to woman or man. The feeling is not to be relied on, however, nor are any of the other moments when motherhood suffuses me with joy, leaves me almost swooning. To rely on them would be to kill them, in fact. Alcohol is much less powerful but much more dependable.

At six p.m. about twenty years from now, I'll be sitting at my dimly lit kitchen table with my beautiful Moroccan glass. The pleasure of wine will have been mixed with the pleasures of motherhood and the pains, the sediment of wanton feeling laid down year after year. I'll still be a mother, but a mother whose children aren't around anymore. I already miss them.

LOUIS BAYARD

A Dad's Affidavit

WHERE DO I BEGIN? How do I apportion the blame? Start at the top.

God

Like many agnostics, I now and then make bargains with Him. They follow the algebraic line of "God, give me *X*, I will give you *Y*." *X* could be any number of things: a kiss, a man, a semi-detached row-house. There's always an *X* on hand; it's the *Y* that doesn't necessarily come when you call. And why should it? You're asking God to fuck with divine plan, the space-time continuum, and in exchange, you will . . . cut out simple carbohydrates? Attend a sunrise Easter service? No, if a speculative being decides to become actual, it's safe to say He will need something mighty in return. It will have to be the thing He most wants and the thing you most *don't* want. That's a *real* man's *Y*.

Six years ago, in the course of awaiting a lab result, I managed to persuade myself (never much of a hypochondriac) that I was dying.

The world gleamed with the dew of my leaving. The birds sang goodbye. And there I was, walking down Pennsylvania Avenue in shorts, threading my way through a horde of Library of Congress tourists, when the words came to me with such force I actually said them out loud.

"God, give me this, and I will become a father."

So strange that, when pushed into a corner, I should have made *that* my *Y*. How could I possibly have believed that God wanted me to be a father? The God of mainstream Christianity? The God of Islam, the God of the Jews? They would have thrown me out of their courtrooms. And frankly, in those days (not so long ago), I wouldn't have got a much better hearing from Secular Humanist God. I was a gay man. I was to acquire my china, I was to tone my external obliques, I was to send gratuitously expensive Christmas gifts to my nephews . . . I was to stay out of the whole parenting franchise.

That, at least, was the rule imparted to the gay men of my generation (coming of age with the AIDS epidemic). The reasoning went something like this: *You're alive. Be glad. Don't push your luck.* And we accepted it, most of us. Figured we'd cut a good bargain. You had only to listen to our dinner-table chatter—pocket doors, chair molding, the vagaries of diets and cleaning women and Rehoboth time-shares—to know that the mention of children would elicit nothing but a baffled silence. You had only to walk into one of our homes and see the *Men's Fitness* magazines and the shell-colored carpets and the Moroccan textiles and the ridiculously breakable vases and decanters and the plunging stone staircases with the open rails to know: *Children will never maraud here*.

So, even five years ago, it was no good pretending God or even other gay men wanted me to be a dad. Why, then, did I continue to posit it as my *Y* through a long, long procession of *X*'s? Maybe because, in my mind, it really *was* the last thing I wanted. And so, it acquired a spiritual dimension all its own and became over time an actual agony—which is to say, a *presence,* nearly physical in the pain it could elicit. And in this way, I was forced to acknowledge that being a father was also the thing I *most* wanted. And with God out of the picture, I was reduced to bargaining with myself.

Men Around the Corner

Since I would not change, the world had to.

It started when these neighbors of ours—a sweet young gay couple named Lance and David—acquired a child. Out of thin air he appeared. A betel-skinned, almond-eyed beauty of a boy fresh from the wilds of Ho Chi Minh City. Name of Ash. During the next couple of years, I don't think I ever spent more than a minute with Ash and his dads, but each moment of witnessing left me stunned by their raging, harrowing, mind-snapping *banality*. They strolled to the park together, they ate food at Mexican restaurants, they sat at outdoor cafés, they crossed in crosswalks. And when I passed Ash riding his tricycle down the sidewalk—one dad on either side—he caught my eye and said, with great calmness: "Bike." He'd said it, probably, to the last stranger; he'd say it to the next. Lance and David and Ash were this radical thing, a *family*, and it shouldn't have come as a surprise but it did. And it put paid to one of the main arguments I had left in my arsenal, which was simply this: it can't happen.

My Brother

This is what my brother Paul did: He got married.

No one in my immediate family has ever had a traditional wedding. My parents were wed by an Icelandic justice of the peace. My brother Chris's ceremony took place on the top floor of a Days Inn in Chicago, overlooking Lake Michigan, and was officiated by a friend who had acquired his ministerial credentials over the Internet. It lasted nine minutes, half of which was me singing "More I Cannot Wish You" from *Guys and Dolls*.

Paul's wedding took place (with three weeks' notice) in an Alameda, California, hotel once frequented by Errol Flynn. The minister was an employee of Weddings, Inc.—his response, upon hearing the marital vows exchanged, was: "Kewl." The bride was heavily pregnant, the bride's uncle strummed Salvadoran tunes during the ceremony, and the bride's mother (whose attendance had been in some doubt) surprised everyone by hiring for the reception a mariachi band, headed by a fierce contralto who kept advancing on our table like an outraged bear.

There *was* an element of the expected in Paul's wedding. Children. Gobs of them. Brown, black, white, blotchy. Spilling punch, getting into fights, turning their dresses into parachutes. Dazed with shyness, hilarious with exhaustion, gibbering and giddy and droopy and still.

It all posed something of a problem for me, because lately I had been avoiding children—for the same reason that Clark Kent avoids kryptonite. They sapped me. Even in the act of ignoring me, they seemed in some way to be pursuing. And it was the same at Paul's wedding. Rivers and rivers of kids, flowing past, not even a glance in my direction . . . and each of them gnarling off a piece of me and swallowing it whole.

Later that night, the wedding behind us, I sat on the beach with my partner, Don, and sobbed like a revival-tent sinner. "I want a child," I said.

It was a moment, I guess, of exquisite helplessness. Despite all I had done to armor myself, despite all the lengths I had gone to avoid children, I could no longer avoid wanting one. There was no running away from that.

"Okay," said Don. "So why are you crying?"

Stendhal

Maybe I didn't mention . . . I was in the middle of a book tour at the time. I did not know what I know now: Book tours are a concentrated immersion in mortality. When you've seen the names of all the authors who were there before you, all the authors who will follow, when you've seen the great slough of despond into which your words are hurled, when you've squinted out at the four, six, ten people (most of them known to you) who have staggered out to find you at an inconvenient address, when you've cleared your throat in preparation for ten minutes of reading and another ten minutes of Q & A, you know: *This is death.*

So I can only conclude that, on that Alameda beach, I was grieving a certain fact: the fact that I would never be sufficient to fill me up. Nothing I did, nothing I was, nothing I had to offer could plug that hole. And this, for some reason, was sad.

But once that point had been passed, once I began actively to ponder life with a child, it was amazing how sufficient—how important—those same things came to be. All I could dwell on was the certainty of losing them. Work and travel and culture and dinner parties and movie nights and all-day bike rides and the unapologetic consumption that goes with being a free adult in a capitalist society . . . it was all slipping away.

And books! Books would slip away, too, wouldn't they? Because once you were a father, you could never again read a grown-up book—not really—not in the way a grown-up was supposed to read a book.

And so, when we started the paperwork for our adoption, I made the decision that, if I did nothing else before our child came, I would read *The Charterhouse of Parma.* Why *this* book out of the entire Western canon should have presented itself I can't say, but it became my defining obsession, and Stendhal occupied a larger part of my consciousness than acquiring nursery furniture or securing day care or mastering Diaper Genies.

I was remembering, you see, what my friend Peg (a woman after my own heart) had told me. "You just have to kill off your old life," she'd said. "And once you do, it's fine."

So be it. So fucking be it. If I could read *The Charterhouse of Parma,* if I could just do that, then I could let that life go. I could let *me* go, with barely a qualm.

And so, as I read, I got the same valedictory feeling I had walking down Pennsylvania Avenue, with the lab results looming. It was farewell time. The words passed into vapor the moment I finished reading them. How sweet they were. How I would miss them. And when I reached the point where Fabrizio, despairing of life, retires to his charterhouse—"in the woods bordering the Po, some two leagues from Sacca"—I felt like I had followed him there and shut the door after us. The book was closed.

Seth

And then Seth opened it again.

He was born (like Ash) in Ho Chi Minh City, née Saigon. Literally dropped into our laps he was. And we were dropped into . . . what? I was going to say fatherhood, except we weren't yet fathers, we were children ourselves, trying to care for another child, trying to get a screaming six-month-old child to fucking goddamn *eat*. Oh, he fought us. Fought hard. And sometimes in the middle of one of his towering rages, we would ladle water or baby mush or custard pudding into his open maw, and he would (of necessity) swallow and then pick up the howl right where he'd left it, and once I looked into his mouth and he was screaming so hard the custard was actually vibrating against his uvula—like a Geiger graph of terror.

We had to leave Seth in Vietnam for forty days while the paperwork went through. And thanks to his foster mom, by the time we returned, he was weaned and plump and no longer howling. I would carry him around the streets of Saigon in a Baby Bjorn, and I would dangle his feet in the hotel swimming pool, and I would point at things through a window and provide their English names, and somewhere along the line, I realized (with some dismay, perhaps) that I had come to depend on him. On his person, I mean—his touch, his smell, the long spidery lashes over his black eyes. Somehow or other, while my guard was down, he had made himself essential.

And that made all the questions of cost and benefit, old life vs. new life, effectively moot. It didn't matter if this was the right choice or wrong choice. It just was. Seth was and would always be, and everything else would have to be refracted through that. Which makes it sound easier than it was. It wasn't easy; it was just the only way.

And now four years have passed. Don and I are in the middle of the paperwork for our second child. This very day, I'm expecting a certificate from my doctor absolving me of major diseases. The weather is unseasonably hot for April. A posse of street robins is pecking around the daffodils (nearly gone). I'm taking my son to Montessori. We're walking down the street to our Subaru, his hand is nonchalantly in mine, and he's asking me what Spiderman eats for breakfast, and God knows what I was thinking, back there with Fabrizio in his charterhouse. God knows what I was thinking on that beach in Alameda. The book keeps going. As though I could have stopped it.

JOAN GOULD

Once More, With Feeling

THE TIME IS A FRIDAY NIGHT IN JANUARY 1967, just before my fortieth birthday. My family of four gathers around the dining-room table for a weekend celebration. Eddy, who has just turned fifteen, is in his last months as a boy, still Eddy rather than Ed or Edward. His hair—formerly blond, darkening now to brown—rises in curls inches above his head, exuberant as if the energy of his intellect, too much to be cooped in an adolescent scalp, explodes into the air above him. In another year, he'll plaster it flat, trying to look like the boys in his private day school.

Kathy, seventeen months younger and in the first blush of approaching maturity, is overtired and frantic because she has such atrocious, such unbelievably gross dandruff; she complains that she has to wash her hair every two days. When her father, Martin, points out that this will only make the dandruff worse (his hair has exactly the same tint of red under the brown as hers, the same widow's peak), she sprinkles salt all over her plate and commands him to look—will he just look?—at the dandruff that has fallen on her food?

Bypassing her complaints, Martin jumps up from the table and hides in the living room behind the couch, calling for our collie, Sailor, to come find him. Sailor, who is beautiful and loving but none too bright, makes an energetic tour of the room, but can't find Martin. This is repeated twice, until Eddy, always the protector, takes Sailor by the ruff and leads him to the hiding place. The next time Martin hides—in the same spot, of course—Sailor finds him, and everyone assures the dog that he's brilliant.

Die Fledermaus is playing on the phonograph, which is what we call the sound system. Kathy gets up and pirouettes around the dining room to prove to us that she ought to go to the culture camp she has just read about rather than the sports camp she hated last summer. Martin, a trial lawyer with a major firm, won a motion in court today. Singing along with Patrice Munsel, he takes a split of champagne from the refrigerator and opens the dining room window to uncork it. Wine gushes forth. "Help yourself, Sailor dear," he calls to the dog, who is in the backyard now, while Eddy puts the silver-foil cork wrapper over his nose, like Pinocchio.

Kathy turns to her brother for advice and comfort about her braces, which will be applied next week. How is it that these two are so devoted to each other, almost like a married couple rather than a pair of siblings? Certainly not because they are close in age. That could just as easily have been a source of conflict between them. Is it because I was sick throughout their childhood, and so they had to depend on one another during the hours when I was prostrate, or worse, when my illness frightened them? Or is it because they are natural complements? He's the academic star, determined to shine in a crowd; she's as self-contained and sufficient as an egg.

It's possible that the two of them will never be so close again. Within a year, my son will have a driver's license and will start thinking about college. My daughter will be a storm-tossed adolescent, with more to worry about than the split ends of her hair that have recently caught her attention. In the more distant future, brother and sister will be separated from one another and from us, their parents, by separate vocations, locations, ambitions, and partners.

My time is nearly over. I can feel the gears shifting, but don't know how to prevent forward motion. What I want is for the four of us to hold still, hold on to the moment, already sliding out from under me, that they are destroying with their antics.

"Will you sit down?" I burst out in anger. "All three of you, sit at the table and eat your dinner." Puddles of gravy have congealed on their plates. The meal is being ignored, my planning for a perfect hour by candlelight destroyed. "Don't you know that leg of lamb is expensive?"

And with that, self-hatred overpowers me. In front of a scene as choreographed as a sitcom, I see that I've failed the family once again. I can't rise to their level of hilarity. I have no taste or talent—not even tolerance—for disorder. Why can't I find a way to release the tenderness I feel?

I know that the evening passed as I describe it, because after dinner, in a bout of nostalgia for the opportunities I've missed—or mishandled—I rush upstairs to write the scene down in my journal before it slips away.

How often in our lives do we make conscious decisions—by which I mean decisions made while we're aware, at the moment, that they are going to affect the rest of our lives in ways we can't possibly know yet, change us into someone we might not have become without them? I can count mine on my fingers. The first was on the blind date when I met Martin and made up my mind, before an hour had passed, that I would marry this man or else marry nobody, now that I knew he existed.

My son and daughter were certainly not "accidents," but neither were they fully conscious decisions. As soon as Martin and I settled down in a New York City apartment and spent the usual honeymoon period of less than a year getting used to one another, we wanted to have a child, just as all our peers did in those Baby Boom years. Giving birth twice in my early twenties was indeed my choice, but it was also the expected thing to do; it was part of the young-married schedule, if the marriage was a happy one, the rent was paid on time, and the gods were kind.

Then why did I decide to have another baby fourteen years later?

During my first two pregnancies, and for a long time afterward, I was seriously ill with what was diagnosed later as Crohn's disease, an inflammatory bowel condition, combined, in my case, with an inability to digest milk. Since no one in those days had heard of a lactase deficiency, my symptoms were aggravated by the fact that my doctors exhorted me to drink lots of milkshakes during pregnancy to make up for my bouts of vomiting and weight loss. Shortly after Kathy's birth, I weighed eighty-nine pounds. I wanted to have another child, but my doctor made it clear that I'd be risking my life to do so.

But fourteen years after Kathy's birth, I was fortified by better health, calcium pills made from oyster shells rather than milk, more leisure, certainly more money. We owned the suburban house that I still live in, close to a small beach with a dock and moorings, a perfect place for a child to play, even if I'd picked it for a different reason. I had always wanted to be a sailor. I knew little about the sport when we moved there, but I'd learn, and eventually I'd buy a boat and become a racing skipper at a time when women didn't own and race sailboats without a husband on board. As for Martin, he refused to go on, much less in, the water, but he was generous enough to allow me what he didn't want for himself.

With a nourishing marriage, and with health, house, money, and beach, why shouldn't Martin and I treat ourselves to the luxury of another child? We must have discussed the idea one night in bed, I must have given Martin some sort of explanation for my desire for a third child, but if so, I don't remember the conversation. In that era, husbands and wives operated in separate spheres. Men provided the means, but women decided the ends. Martin worked five or six days and several nights a week to support us. I was in charge of the house and children, along with my freelance writing and boating. If I chose to have another child, if I chose to go out sailing in stormy weather, that was all right with him, so long as I earned the money to buy and maintain the boat, and so long as the children were well cared for.

But he must have realized I was suffering from a hunger that he could help satisfy, an appetite that he shared. This was a man who adored his children. Every Saturday morning, he took Kathy and Eddy to the toy store, over my puritanical objections that he was spoiling them, and bought them each a toy, plus a rawhide bone for

Sailor, who didn't like bones but pranced around the house with the new one in his mouth, to show off how well loved he was.

Months later, when neighbors stared at my belly to make sure that I hadn't simply put on weight, Martin put an end to their speculation: "We didn't trust our children to do right by us," he joked. "So we decided to produce our own grandchild." There was truth mixed with pride in what he said. He was forty-eight years old at the time, but to be the parent of a small child was to be young and potent.

In my carpool groups, the mothers who had children about the same age as mine were laying life plans of a different sort. We were the first group to turn forty while the waves of the feminist revolution rolled under us, lifting us up but washing away our footing. Our mothers' generation never dreamed of working—or was never allowed to work—unless the family was in such need of money that the situation had to be made public. Our daughters' generation never dreamed of not working, except that they scorned jobs. They planned to have careers, for which they'd prepare at graduate school, and then later, once they were well established professionally, they'd make time for marriage and motherhood. We were the in-betweens: college-educated women with degrees that were out of date, tennis players with teenage children who would soon leave us alone in the house and unemployed.

Two of my fellow carpoolers went back to college for degrees in education. One acquired a master's degree in sociology and held a high position in the state prison system for women. Another worked at a mental health clinic. I was the holdout, or dropout, who refused to walk through the door newly opened for us.

Even to myself, I couldn't rationally explain my motive. If another child was to be my midlife career, that wasn't solely because I found motherhood more meaningful than anything else I had done (though that was true), but because the converse was also true. The shameful fact was that I didn't think I had been a very good mother. I had been exhausted and irritable much of the time, which could be attributed to my illness, to be sure, but wasn't my children's fault. I believed that if my teenage son and daughter were so supportive of one another, that must be proof that their mother had failed them in their early years. When Kathy was two years old and Eddy three, he was the one who had to do the back buttons on her sundress on mornings when I couldn't get myself out of bed, just as she poured out his breakfast cereal a year or two later. Then there was the day I can scarcely acknowledge, when I had to leave the two of them, still toddlers, clasping hands outside the door of the employees' toilet in a New York City supermarket, with instructions not to let go of each other under any circumstances. Mommy would be out in a minute.

But none of that could be helped. There were other rifts in my conscience. Why had there been so many hours when I was with my children and yet not entirely *with* them? Why couldn't I be a mother and a person at the same time, if I could be a person and still carry another individual inside me?

As a writer, I felt the need for silence, solitude, my own thoughts welling up unbidden, the sound of my self rather than my responses to others. The thought interrupted never returns, I'd discovered. Or at least I never return to it. I'd found another way to avoid distraction: I'm nakedly competitive, and for that reason I was a skipper. Singleness of intent is what I find on a racing boat, the determination to win, with no interruptions to my concentration.

Solitude and silence, or else the tumult of a race. My pure self either way, in passive or active mode, rather than my mothering self. But there was something in the middle I'd brushed aside. Feeling. Tenderness and gratitude, the emotions that arise from not just looking after children but looking *at* them, staring into the moment that is already carrying them away, with the realization that they will never be quite this young again.

What I wanted was simply, if improbably, to rewind the tape, lead my life as a mother all over again—with more joy, more vigor than had been available in my younger years, and free from the fear that if I didn't do everything perfectly—no license to run out of milk or leave the beds unmade—my children and home would be taken away from me. Out on Long Island Sound, I could prove my competence by battling other sailboats. While I worked alone at my typewriter, I was in a world of my own making, but in my womanly roles, I had no idea how I was supposed to measure success or failure. Now I wanted another fifteen years as the mother of a young child, even as the middle-aged mother of a young child, but this time I was going to be different. I was going to be openhearted, good-humored, buoyed by my awareness of the miracle going on first inside my body and after that inside my family.

In 1999, with the help of fertility treatments, playwright Wendy Wasserstein gave birth to her first child at the age of forty-eight. My next-door neighbor, a college professor, was forty-six when she gave birth recently to twin boys after six rounds of in vitro fertilization. What was thought of as foolhardy, if not impossible, in my day is now considered the triumph of will over statistics. I was lucky enough to be given what I wanted without cost.

But how was I going to feel in the years ahead, while arranging carpools and playdates with mothers closer to my daughter's age than mine? As it turned out, carpools in the late 1960s and 1970s were different from those in the 1950s. There was a grandmother in our group whose divorced daughter had vanished, leaving her with a grandson to raise. A woman new in the neighborhood had a teenage daughter but gave birth around the same time I did, to a boy who became my youngest child's best friend. Later, I found out that there had been a son in between, who had been killed by lightning. Age didn't seem to matter as much as it had in the past. Measles and mumps vaccines were new in this baby-raising world, as were paper diapers and sippy cups. But I learned that women who have children the same age are of the same vintage, no matter what their chronological age.

I have always—three times, that is—convinced myself that I'm aware of the moment of conception. Staring at the ceiling, I have a momentary feeling of fullness, and the deed is done. But months of self-doubt follow, in which I wonder if I'll have the energy day and night, seven days a week, for what I've set out so irrevocably to do.

In my early twenties, I had been fearless—maybe biology had spoken so loud inside me that I hadn't heard my own thoughts—and everything had worked out fine. But in middle age, my decision was aberrational, and, in the dark, I was mocked by anxieties that hovered above my bed. Was I fooling myself? Was I having a baby because I couldn't think of anything else I wanted to do? Motherhood was a job I couldn't quit, after all, if I found it tiring or boring, the way I could quit law school or a teaching post I didn't like.

With daylight, my assurance returned. A baby was more than what I wanted; it was what I had to have in order to become the mother I thought I could be. Ironically, the only pregnancy of the three that represented a defiant stand against what the community expected of me was the one that my peers assumed was an accident, not only unwise and unheard of but dangerous.

"Opinion in the community is divided," one friend said. "Fifty percent of the people think you're crazy, and the other half think Martin is." She herself had a degree in architecture and was working as a city planner. "I would never have patience at this age," said another woman, not without admiration in her voice. "The diapers, the feeding. All that mess." "Aren't you afraid of birth defects?" a third asked, justifiably, in those days before amniocentesis, but the truth was that I was neither afraid nor unafraid. Probably the statistics weren't common knowledge yet, but in any case the thought never crossed my mind. Or was barred from it. I never doubted that my child would be born normal, especially since this was the first healthy, even joyful, milkless pregnancy I'd had.

If I appeared ludicrous to the neighbors with my middle-aged face above my maternity clothes, that didn't bother me. I'd looked worse for years, whenever I went sailing in foul weather. And if my course of conduct seemed peculiar, that was just part of the package for a woman who raced in any weather with two or more men aboard, who were other women's sons or husbands, as crew. The suburban gossips had long since stopped gossiping about me.

For me, the real question was whether I would succeed in being more attentive than I'd been in the past. As proof of my intention, I offer the fact that I sailed into the eighth month of pregnancy, and then, as soon as the baby was born, sold the boat and didn't sail again for five years.

• • •

The scene shifts. Once again, we're eating, but we're in the kitchen now, a senior couple and a junior couple gathered around a messy table with the playpen at our side, while the baby consumes anything offered to him—potato sticks, grape juice, milk from a glass—with happy abandon.

Martin gives the baby a bit of steak. David chews with relish, but Edward goes into a panic that he may choke, until Martin takes the meat away in the interest of safety and gives him a sip of Coke instead. David guzzles it, enchanted—all of us laughing, because I had assured them from my vast experience that babies hate carbonated drinks. Getting up from his chair, Martin jigs up and down wildly, singing "*My mother gave me a nickel . . . to buy a pickle . . .*" At that, David begins to sing and jig in his pen, still holding on to the railing so that he won't fall, and Martin's heart flies out to him—I can almost feel the air part—in a violent rush of love, of admiration, too, for one so little and gallant that he dares to hold his own against the claims of four tumultuous giants. For a year now, Martin has seen his son, but this is the first time he truly *sees* him.

I see Martin seeing his son, and at the same time I see Kathy and Edward—see them with stereopticon vision, not only as David's siblings and our children but as the parents they are going to become someday, more permissive and enthusiastic than I was and yet protective; better parents than I've been. But if they're better, that's in part because of what they're experiencing right now, what I've given them.

David was born two days before Edward's sixteenth birthday. Edward drove the two of us home from the hospital while Martin was in his office. Kathy asks for permission to take the baby along in his stroller when she and her friends walk to the beach. "When you have your firstborn, it'll be your second child," I tell this junior set of parents, and, many years later, after becoming a mother herself, Kathy confirmed that I was right. "It was living with David that convinced me I wanted to marry and have a child," she said.

In my memories of David's early childhood, I see him trying to fly.

He learned to walk and run, and then, when he could navigate as we do, he was determined to soar. He used to jump—or, rather, hurl himself—down the flight of stairs outside his bedroom, vigorously flailing his arms and terrifying me. Or he would run down the dirt road toward the beach as fast as he could go, his arms flapping like the wings of a gull, as if convinced that he would lift off from the ground if only he could get up enough speed. The image is comforting. It strikes me that a child who's convinced that he can be airborne hasn't had his spirit clipped.

During the first days after David was born, I was enveloped in a golden haze like a Byzantine icon against a background of beaten gold leaf, and some memory of that golden time stayed with me after I left the hospital, clung to his skin and mine.

I changed. There was no doubt in my mind about that, but the change was not what I expected. I was more openhearted, yes, kinder and more tolerant, but the increased compassion was not so much toward my child (I think now that I had always been kind in that direction) as toward myself. I was no longer so critical of my performance as a mother, or full of fear. I no longer assumed control over every element of the household. And so I began to forgive myself for having flaws.

I also had a lot of help the third time around. For the first two years of David's life, I had his siblings, supplying whatever I lacked in energy and playfulness (Martin never lacked either of these), and for the next two, I still had Kathy in the house before she left for college. But what really changed me was the perception of time passing, which is a gift, or curse that comes to us in middle age—this child too, like the others, outgrowing me, not needing me, growing as strong as I was, and then stronger while I dwindled. There would be another baby in the family sooner or later, but it wouldn't be mine. It would be Kathy's or Edward's turn.

The time frame shifts again. The baby has just turned eleven. His big brother is married, his sister is about to be married—having found her brother's equal in his law school roommate—but the wedding has had to be postponed. Their father is in a hospital in New York City, dying of cancer at age fifty-nine. Some will say this proves the case against having a child late in life.

"I want to go to the hospital and see him," said David on the December night when I told him that, contrary to my assurances, his father couldn't last much longer. Later, David would blame me for not having revealed the prognosis sooner because of my wish to protect him. Only when years had passed did I recognize the deeper truth that lay beneath my silence: I couldn't tell him what I didn't know myself and wasn't ready to know, that there was only one way this hospital stay could end. I needed the normalcy, the distraction that my young son offered when I came home from the hospital each evening. I was protecting myself, at least as much as him.

"You can't visit Dad now that you know," I answered. "You couldn't act natural and joke with him any longer."

"Oh, yes I can," he said. He never had the chance. Martin died at half-past three that morning, according to the death certificate. Between three and four, that had always been the darkest hour for him as an insomniac, the hour when it was too late to go into the bathroom and take a sleeping pill but too long to wait until morning, when the furnace would begin to warm the room in winter or the underground sprinklers turn on with a clang in summer.

No one told me to stay with my husband in the hospital that night, just as no one had told me to bring his younger son for a visit a few days earlier. If Martin woke up for the event and coughed, or cried out, or said he was in pain, or left a message for me or his children, or didn't leave a message because his lungs were filling with fluid, I never heard about it. All I knew was that he died by drowning, this man who would never set foot in a boat.

Grief, I discovered, isn't steady weather. It comes in gusts like northwest winds. I hid in the bathroom and buried my face in the towels hanging on the rack so that David wouldn't see or hear me. For all I know, he may have done the same.

Instead of facing the fact, I invented a game. My husband wasn't dead. He was just out of town on a long business trip—no, he was the captain of an old-time whaling ship, who might spend two or three years circling the oceans until the hold was full of oil and ivory. As his wife, I was expected to mind the child and home he had left behind, and the family finances as well, so he would find everything in good order when he returned. Soon I discovered that a child—my child anyway, perhaps every child—has a function we don't expect: He's an anchor to windward, set out in stormy weather, that keeps us from dashing ourselves onto the rocks or piling up on the beach. Instead of allowing us to be blown off the mooring, he holds us in place, heading into the wind while we wait for fairer weather to come.

The weather system blew over, as it always does. On the day of Martin's funeral I got a letter from a publisher, telling me that my first book had been accepted for publication. In the next few years, Kathy married and had a child. Edward married, divorced, remarried, and had a child. David entered high school. My personal life opened up in unexpected ways. I had assumed that I would be alone when David went to college, but after four years of widowhood, I went to a high school reunion and met a former classmate, who has been my partner for more than twenty years now.

Would I have decided to have a baby if I had known that Martin would die in his fifties? The other way around, would I have survived as well without a child who demanded strength from me rather than weakness? How can I tell? At the important junctures in our lives, when we fall in love, marry, conceive a child, pick a vocation, we are inspired by our gut if we are lucky, rather than our brain. We choose a direction, with no idea where it will take us or how we will change along the way.

I can only say that my choice of direction—which is not the same as destination—was as conscious as any I have made in my life. As the philosopher Martin Buber once put it, "All journeys have secret destinations of which the traveler is unaware."

AMY RICHARDS

Triple Threat

I WAS TOO NERVOUS TO LOOK AT THE RESULTS of the pregnancy test, so I instructed my boyfriend, Peter, to go into the bathroom to inspect the strip. During the five minutes we'd been waiting, I had played with the two possible outcomes in my head. If the test was negative, I would worry that my body didn't work, that this might be the first of many disappointments. But if a positive result waited behind the bathroom door, then what? Were we really ready for parenthood?

Peter returned from the bathroom with a confused look on his face. Rather than a neat one- or two-stripe result, the entire swab was bathed in pink. We quickly called the 800 number listed on the back of the box for help, but the customer service representative was just as perplexed as we were. She suggested we get another test.

The second test, which we took later that evening, had two unmistakable lines. I was pregnant. We were surprised and entirely disbelieving. We'd only recently agreed that I would go off birth control pills, and had been pretty casual about the whole thing. No ovulation kits, no spreading the word that we were "trying," no rush to get married—we just wanted to see what happened. While the reality of having a child was daunting, I was also giddy—our secret experiment had worked! Like a naive adolescent, I couldn't believe that it really only took one time.

A few weeks later, Peter and I sat in my doctor's examining room as she reviewed the general rules of pregnancy—no caffeine, no alcohol, no sushi, and yes, sex was fine. Given my age, thirty-three, and the fact that I was healthy and athletic, my doctor said she was confident that I would have a low-risk pregnancy. The sonogram was the last step.

The lights in the room were dimmed, and I lay back on the white, crinkly paper, eternally confused about whether the opening on those blue medical gowns goes in the front or in the back. Peter sat beside me, as the doctor wiggled a wand inside me in search of the fetus. She finally focused in on something, and moved closer to the screen. In a detached, professional tone, she said, "Are you sure you've never taken fertility drugs?"

My first thought was that the baby was horribly deformed. I took a few seconds to respond and then emphatically said, "Yes, I'm positive. Why?"

"Because you're pregnant with triplets."

We all sat in silence for a few seconds. I thought back to when I was pregnant at nineteen and learned, when I went in for an abortion, that I was pregnant with twins. What were the odds, I wondered, of having twins and then triplets? How could one person possibly have multi-multiples? The stick bathed in pink suddenly made sense.

My doctor finally broke the silence and said she was just as shocked as we were. She pointed to a dark mass on the screen, but it just looked like a blob to Peter and me. She told me that carrying triplets, because of the added stress to the mother and the fetuses, often mandated extended bed rest. I thought about my work as an author and feminist activist, which required that I travel to college campuses all over the country. How would I make a living from my bed? I listened to my doctor and remained calm as I tried to get my head around the news. Finally, I asked her what I'd been wondering since the moment she had said the word "triplets."

"Can I get rid of one or some of them?" I said, half in jest, since I didn't think it might actually be possible. Yes, my doctor said, there was a procedure called selective reduction. She didn't have many details, and she wasn't certain if it was possible to reduce more than one fetus, but she promised I would have options. Even though I had worked closely with reproductive health organizations my entire adult life, I had never heard of selective reduction.

I always imagined that the moment I found out I was going to be a mother would be the instant that I would truly and finally become an adult. And yet, in that moment, I felt more like a child than ever. I wanted my doctor to tell me what to do. I wanted her to tell me that selective reduction was what I *should* do. But all she could do was give me phone numbers: one for an obstetrician who specialized in high-risk births, one for an abortionist, and one for an expert in selective reduction.

In certain moments, I thought about aborting all three fetuses and starting over. When Peter suggested that we consider having all of the babies, I snapped, "It's not your body." Even though I knew he was broaching this possibility because he wanted me to feel supported, for us to consider all of our options, I just couldn't imagine life with three babies. I was an only child raised by a single mother. It was hard enough for me to imagine having three children spread across my child-bearing years, let alone three all at once. I wasn't living large, but I had become accustomed to my East Village apartment, to working, to having an occasional dinner out. I liked to travel, exercise, see my friends. I also had a vision of what I wanted for my children—summer camp, a home comfortable enough so their friends could visit, trips to other states and countries, swimming lessons. All of this seemed impossible for me to provide for three children—especially since at the time we were making this decision, Peter was out of work. When I imagined life with three babies, I thought about them all having to share one thing—three of them tucked into a crib, three of them smushed into a stroller—since there wasn't room for more than one of anything in our tiny fifth-floor walk-up.

While I waited for an appointment with the selective reduction specialist, I started doing my own research. I scoured the Internet for information, I called colleagues and friends who worked on women's health issues, and I obsessively read medical journals to find out everything I could about the procedure. I learned that selective reduction involved a shot of potassium chloride into the heart of one or more fetuses at around twelve weeks. I learned that many doctors dissuade women from carrying more than two children, because of the increased danger to mother and child. I learned that thousands of selective reductions are performed each year—the procedure has become a lot more common in the last decade, with the rise in infertility treatments and the attendant multiple births—and that most involve reducing from four fetuses to two. I learned that in selective reduction, the body simply absorbs the fetal matter, unlike an abortion, in which the fetus is removed and you can literally count limbs. I learned that the procedure is performed in a hospital, which frees a patient from the potential threats that may accompany a trip to an abortion clinic. And, most interesting, I learned that women who opt for selective reduction do not refer to the procedure as an abortion. Because they usually do it under doctor's orders, many don't consider it a "choice," and therefore call it by its less controversial name.

Not surprisingly, given my job as a professional feminist, my pregnancy quickly became both a personal dilemma and a political investigation. Why, I wondered, was this fairly common procedure so rarely talked about? Why was abortion so politically charged while selective reduction—essentially the same thing but with a different name—neither protested nor politicized? Is the euphemistic procedure "selective reduction" somehow more palatable because women who have them still become mothers? I realized that whatever you call it, the decision to abort or selectively reduce is ultimately about autonomy. My political inquiries allowed me to distance myself from the personal ramifications of my decision. I poured my energy into my research and was relieved to discover that many women before me had made similar choices.

At my first appointment with the specialist, I found out that two of three fetuses were identical twins. According to the doctor, this brought on an entirely new set of complications—two babies dependent on the same placenta, and a greater potential for the umbilical cord to suffocate one of the children, because they were sharing a small space usually reserved for one. Even if I hadn't wanted to reduce from three to one, my doctor would have recommended it. In some way, this made me feel much more comfortable with my decision. My personal preference for one child, and the quality of life that I wanted for my family, did not seem like a good enough reason to abort two fetuses. Even though I knew this was the decision that I wanted to make, publicly I was still uncomfortable with my choice. I felt like I needed to have some way of making it a necessity, not just a selfish decision. So I put myself in the same category with women who went to great lengths to get pregnant using technology. They were creating their ideal in much the same way I wanted to create mine. They desperately wanted to be mothers; I didn't want to be the mother of three.

I had the reduction during the eleventh week of my pregnancy. A needle was inserted into my belly while I lay motionless on an examining table, fully clothed, with my pants pulled down enough to expose my stomach. It was essentially painless. I was out of the hospital in twenty minutes and then had to stay home and rest for a day.

As my pregnancy proceeded, I kept waiting for something horrible to happen. Karmically, I felt like the decision to reduce would come back to haunt me. I worried about miscarrying, about unforeseen complications, about giving birth to a sad, miserable child. But none of my fears came to pass. I now have an active, happy twenty-month-old son. I devote a part of each day to doing something with him—an art class, a play date, a trip to a museum. And I never think about the fact that it's only the two of us on these outings, and not the four of us. I have no regrets. I will likely have to contend with these choices again. I want another child, and as my doctor has told me, my eggs "just like to split." What will I do next time? I really have no idea.

ALISA VALDES-RODRIGUEZ

Diagnosis: Broken

PATRICK AND I SAT IN OUR GAUDY PURPLE Plymouth Neon in the parking lot of the Kaiser Permanente health center and debated.

"It's going to be a girl," he said.

"No," I said, sicker than I'd ever been in my life, a walking fountain of vomit with lower-back pain. "Only a boy could make someone this miserable."

Twenty minutes later, the ultrasound technician pointed to an unmistakable little appendage, and Patrick shook his head in disbelief.

"It's a boy," said the technician.

That night, we decided to name our son Alexander, and we sat up imagining all the things he might be. He might be a poet, like one of his grandmothers, or a shopaholic, like the other one. Maybe he'd be a professor, like his short grandpa, or an electrician, like his tall one. He might play baseball, like his dad, or saxophone, like his mom. We had so many grand ideas, so many great plans. He would be the perfect expression of us. A boy. A beautiful baby boy.

What we never talked about, what most parents never think to talk about during the months after they've made the decision to have a child but before that decision is a living, breathing reality, is that your child, the one you are projecting all of your dreams onto, might be broken. It simply never crossed our minds that Alexander Rodriguez could come out wrong.

Alex did come out *big*. I, of course, was acutely aware of all nine pounds, eleven ounces as he popped into the world. He hadn't dropped, even at the onset of labor, so he was cut from me as I lay on an operating table that looked like a black vinyl padded crucifix. The blue sheet hid from view the things that I could hear—my belly being sliced open, the doctors plunging their hands into my gigantic ball of uterus, the yanking of a huge bloody boy into the light.

"He's a boy all right," one of the doctors joked.

"We've got a big 'un here," said the other.

A nurse wiped my son, and brought him to me, placing him on my chest as they finished stapling my flapping tent of a stomach. I trembled uncontrollably, partly from shock and partly because *there he was.* I had waited so long to meet Alex, had imagined his face from the ghostly ultrasound outlines, and now he was here. And he was beautiful, with his father's hands, staring right into my eyes with his own squinty ones.

Patrick, at my side, gasped in wonder, and he cried. We both cried. I held the tiny fingers, and they curled instinctively around my big thumb.

"Welcome to the world, Alexander," I said. "I'm your mommy. This is your daddy. We're here to take care of you and love you forever and ever."

The clichés, and the frequency with which they are uttered, are almost sickening. "Your life will change forever." "You'll never feel a love deeper than what you feel for your children." And from my father, "Once your children are born, you will never be able to remember life without them." I'd heard all these pearls of wisdom, and thought they were trite and annoying—until my son was born. Then they all became true, in a bright, sleepless, Technicolor way.

Life changed in the normal ways. I didn't sleep for two years. My breasts ached and dropped a couple of centimeters from having been turned into Alexander's personal dairy. I was floppy in places I'd formerly been taut. Sex, particularly the wild spontaneous kind, became a distant memory. And I was happier and crankier than I'd ever been in my life.

Because we'd never had a kid, and didn't know any kids, we read all the developmental books. And we cheered when Alex reached most of his milestones early. He turned over early. Sat up early. And talked early.

Way early.

The average kid can speak one word by his first birthday. Alex, when he was seven months old, stared up at the ceiling in our apartment and said, clearly, "fan." Five months later, he had a documented vocabulary of sixty-five words, and was speaking in sentences. At his one-year checkup, the doctor asked us if Alex was trying to say "mama" or "dada" yet. Patrick and I exchanged smug smiles and reported that Alex had been saying "mommy" and "daddy" for months, and could furthermore name colors, shapes, and animals accurately.

"He's doing sentences," I said.

The doctor stared at me as if I were insane. Taking his cue, Alex grabbed the stethoscope and asked, "Mommy, what is this?"

The doctor stared at us, then grinned and said, simply, "There's nothing special you need to do. Just love him, nurture him, and let his genes take him where he's going to go."

Around this time, I signed Alex up for a playgroup. Even though I was writing and publishing novels, I joined something called The Moms Club, a group for stay-at-home mothers. I was staying at home. I just wasn't a housewife. But the women in the group didn't need to know that. They also didn't need to know I did not share their politics (Bush), their religion (Jesus), or their taste in cars (minivans). They just needed to bring their kids over to play.

It was fun for a while. But after six months, playgroup became awkward. All of the other kids were walking quite well, having started at about ten or twelve months. At eighteen months, Alex was still crawling. The other kids interacted more with their parents and with the other children, while Alex liked to stare at shapes and patterns. The other boys were running and jumping, but didn't talk much. As Logan and Walker playacted at being the superheroes they'd seen in Disney movies, Alex sat quietly in a corner and made perfect dinosaur replicas out of tiny Legos meant for much older kids.

While the other parents struggled to get their kids to pay attention to books for five seconds or more, Alex focused like a grown-up, and memorized two hundred complicated dinosaur names. By age two, Alex was stacking his wooden blocks about thirty high, climbing onto the coffee table to stack them even higher, his brow furrowed in concentration. The playgroup friends watched, but could not do the same thing. My mother, observing his seriousness and remembering how goofy my brother and I had been at that age, suggested my son might be clinically depressed.

One day, after I'd worried myself sick that Alex, then eighteen months old, would never walk, he pulled himself up to the windowsill in our home, turned toward me, and walked. Perfectly. He did not waver. He did not stumble. He did not fall down. He was speaking in very coherent sentences by then, so I asked him why he'd waited so long to try walking.

"Alex doesn't like to fall," he said.

Alex didn't like to fall, or make messes, or do things for himself if we could do them for him. Or that's what we thought, as we hovered and doted on him, joking that our boy had two full-time slaves, because both parents were always home and always jumping at the slightest whimper to fix all that we thought was uncomfortable, difficult, or unpleasant in Alex's life. We let him sleep in our bed, in between us. We spoon-fed him past the point at which he was able to feed himself. We never let him cry more than a second or two before changing our actions to please him. In a way, it seemed like our own lives had been subsumed by his, but we didn't mind—we'd both been kids who had been left alone too much, and we weren't about to make that mistake with Alex.

When Alex was going on three, he became obsessed with sprinklers and sprinkler systems. He had no interest in any other kinds of toys. It was sprinklers, sprinkler parts, and, eventually, the automatic system in our yard, which we allowed Alex to rip apart, trying to find out where the water was coming from. So what if it was a costly repair? We were fostering our son's creativity, right? And we, being the type of parents who felt discipline was a word used only by stern born-again types, let him do just about anything he wanted. We indulged our little child, giving him as many sprinklers as he desired. We never said no. To anything.

Part of this was because Patrick and I had both grown up poor, and we liked being able to give Alex things we never had. But another part of it was that Alex wasn't the kind of child you could easily say no to. It's not that he was too cute, nothing like that. Rather, he was a demon. If you said no, or if you didn't do what he wanted, Alex threw the most intense tantrums I had ever seen. My mother assured me I'd been the same way, which inspired me to continue saying yes to his wants and demands; I didn't want to be like my mother, who had raised me to believe I was difficult and ruined her life.

Back at playgroup, the other children had begun to hold hands and talk to one another. Alex sat away from them, playing with sprinkler parts. He spoke, but usually just to describe to himself what the people around him were doing. "That girl has a pink flower on her shirt," he'd say while the other kids said hello to the girl. "She's playing in the mud. Her jeans have pointy pockets with rhinestones. She has a watering can that is purple with a lime green spout." The other moms stared at me like I was the caretaker of some kind of alien born of a pod. Their children didn't know "jeans," much less "rhinestones," or "green," much less "lime green." Alex didn't interact the way other kids did.

We enrolled Alex in a preschool, hoping to increase his social skills and allow us to have more time for work. In school, as at playgroup, he stayed to himself. When other kids played tag and hide-and-seek, Alex built a model of the Toronto skyline out of Legos. His teacher at the time, a wonderful woman named Sara, found him adorably different, and left it at that. She even bought a special Lego table just for Alex.

By age three, Alex was speaking well but strangely. All of his classmates were referring to themselves in the first person. Alex continued to refer to himself in the second person. If you said "Hi Alex, how are you?" he would reply, "You're fine," meaning that he, Alex, was fine. If he wanted me to pick him up, he'd say, "You want me to pick you up." My husband and I were starting to become very worried. A former journalist, I went into reporter mode and started exhaustively searching the Internet for answers to my son's behavior. All signs pointed to autism. High-functioning autistic kids speak of themselves in the second person; some continue to call themselves "you" their entire lives. They prefer things like fans and sprinklers to "real" toys. They don't interact with other kids. Many of them sound like "little professors," because they speak so well.

The only catch, the literature said, was that they didn't know what they were saying. This means that autistic children, for however smart they sound, are often just acting like little digital recorders, repeating everything that they have heard, often with the exact intonation of the person who said it first.

When I read this, I felt like my heart was slowly being threaded through with pins, and I could not breathe. Alex didn't know what he was saying? For days, I staggered around in a funk, sick to my stomach, terrified, wondering what it meant to be a parent of a child who might not know you are there, or, if he did know, might not care. All the dreams of the kid who smiles at you from the stage at his school play, all of those images of a reciprocally loving relationship flew away like scared pigeons. Could love be something given only in one direction? Could you give and give to a person, even your own child, if they might never have the biological ability to give back in the traditional ways? It was a level of existential anguish I was unaware existed.

I promptly stopped researching, and hoped it wasn't true.

Alex was in a new preschool class by now, and his teacher, Wendy, was concerned. When the class transitioned from one activity to another, Alex flapped his hands, a classic sign of autism. He didn't talk to anyone in particular, but had a running list of words and phrases he felt were "bad," such as "hug," "hot," "self," "have to," and, most horribly, "love." He would recite them all day long, a ritual. If someone else used one of the words—and there were about forty or fifty of them—

Alex would say, "That's a bad word, the one that Mommy said." He said this in exactly the same way, over and over. "That's a bad word, the word that Daddy said." "That's a bad word, the word that Wendy said."

Wendy suggested we have him screened for autism.

Autism. There it was again.

My heart began to break. Not poetic breakage, either. Not the breaking of the lovelorn. It was the mashed-underfoot-by-soccer-cleats kind of breaking, a destruction that left me breathless, drowning, gulping at air I couldn't find anymore. The boy I'd thought was simply brilliant was possibly autistic, meaning biologically incapable of empathy, profoundly unable to communicate like "normal" people. Where I'd once loved Alex's obsessions, thinking they were the sign of a focused mind, I now saw them as a defect. Alex's furrowed brow no longer made him look studious; now he looked demented. Where I used to think he kicked the cat to get attention, I now saw a child who did not understand at the most basic level that something else could have feelings. Where I'd once celebrated his vocabulary and ingenuity, I found myself cringing when he spoke, growing angry at him for being so different.

I finally took Alex to be tested. We went to the Southwestern Autism Network, a clinic housed in a dull gray block of building in an industrial part of Albuquerque, near a run-down Denny's. Inside, it was like a slightly dirty doctor's office. The waiting room was a square, with square furniture stacked squarely against the walls. One corner had a box, filled with broken plastic toys and headless dolls. A man came and introduced himself as a pediatric speech-and-language pathologist and took Alex away for observation, leaving me with an intake counselor to fill out a lengthy questionnaire.

After four hours of observation, which consisted of having Alex play with the pathologist in a dank room filled with more half-broken toys, the pathologist and a developmental pediatrician sat down with me. They looked sad, like abandoned dogs, as they told me that my son was indeed autistic. Alex was in the room, listening, as they informed me that he likely lacked a "theory of mind," which was why he didn't understand the first person. What this meant, they said, was that Alex was incapable of differentiating himself from the rest of the world. They said it was doubtful that Alex would ever spontaneously tell me that he loved me. He would forever be handicapped in communication, but with medication and intervention, he might be able to lead a productive life.

I cried all the way home. Alex, in the backseat, asked me what I was doing. I said crying. He wanted to know why. I said because I was sad. Why was Mommy sad, he wanted to know. Patrick was helping his parents with a home improvement project. I called him on my cell phone from the car and told him the news. He was totally silent. I think he'd been holding out, thinking my paranoia was the only thing making our son look autistic. After a minute, he spoke. "Why are you so sad? If you're grieving, it's for a boy we never had. You better learn to love Alex for who he is, instead of hating him for not being what you want him to be. He's still Alex, and Alex is great." I agreed, in theory. But in practice? In practice, I wanted to die.

When I got home, there was an e-mail waiting for me, from the intake counselor who'd helped me fill out the forms, a woman who had shared with me the fact that both of her children had autism.

"You'll feel like you're in mourning," she wrote, understanding more, I thought at the time, than my husband. "You will feel like the child you thought you had has died. He might get worse. But you will come to the realization that if Alex grows up to fold pizza boxes for a living, he'll be making a contribution."

I wrote back, thanking her for understanding, and asked her how she'd come to terms with knowing that love would be a one-way proposition in parenting an emotionally unresponsive, autistic child. She assured me that autistic people love, but in ways non-autistic people have to learn to see. She offered to hook me up with support groups. I felt briefly comforted.

As the news of Alex's condition started to sink in, I began to slip. I cried constantly, only stopping when I finally fell asleep. My body broke out in a rash that three different doctors couldn't diagnose. I shook uncontrollably. My mind felt poisoned by my own thoughts: What could I do with a kid like this? A child who understood nothing, loved no one? A child who might never have a first date, who might never go to college, much less even finish high school? The person I loved most in the world was trapped, weird, broken, wrong, horrible. And I could not fix it.

One night, when Alex and Patrick were asleep, I got a knife. I took it into the empty bathtub with me, hoping to end the suffering—at least my suffering—in the neatest way possible. I cut my skin. Not enough to draw blood, but enough to hurt. The moon was full, and light poured through the two skylights in the bathroom. I stared at the wimpy wound, and cried. I had loved Alex more deeply than I'd ever loved anyone. It was my job to protect him, and I felt as if he was trapped behind a wall somewhere, and I wasn't able to save him. I felt like a terrible failure. I didn't think I was strong enough to care for a person who did not care back. I was facing a life of pain and grief, and it felt too heavy to bear.

I wished I could cut harder. Alex wouldn't know the difference, I reasoned, because he didn't know I was there. But I couldn't. I was afraid of the pain.

The next night, I got into the car with the intention of driving myself off the side of Sandia Mountain. Patrick knew this, and stood behind the car as I wept and raged, risking his life to save mine.

"If you go, it's over my dead body," he said. I let the cliché slip. Clichés didn't bother me anymore. I was a cliché. I was worse than a cliché. I was, to use a cliché, the walking dead.

We sat in the car most of the night, crying for our child, wondering how we were going to be parents to a child who would forever call himself "you."

Alex had been diagnosed with a form of autism called Asperger's syndrome. Named for a Viennese psychologist who discovered the disorder, Asperger's is marked by severe social impairment, and restrictive and repetitive patterns of behavior. People with Asperger's differ from other autistics in that they have normal or above-normal intelligence, and seemingly normal language development. What they're not able to do is read facial expressions or social cues. They are the kind of people who get bullied the most in school, the misfits and eccentrics, the ones who speak robotically and louder than necessary.

I devoted myself to learning as much as I could about the condition, obsessing over articles and information so much that I even joked that my ability to focus so single-mindedly was sort of Asperger-ish in itself. Some very famous people, I discovered, were possibly afflicted with the syndrome, including Bill Gates and Al Gore. And with intervention, Aspies, as they are affectionately called, could have productive lives.

Alex, who had been listening to my conversations with friends and family about all of this, saw me concentrating on my research one day and said, "Mommy, what does 'cheap' mean?" I said "cheap" was something inexpensive or something that wasn't put together well. He frowned and pointed at himself. "Alex is a cheap little boy," he said.

I stared in astonishment at my child. How had he gotten this idea? From my constant conversations about the subject? Was he listening so closely to what I was saying? And if he was, wouldn't he be doing something considered impossible for most Aspies, in that he was looking at himself from my point of view? The old Alex appeared for a moment, the one I'd thought was just plain old smart and quirky. His shoulders dropped, and he watched his feet. He had already internalized this sense of brokenness. In that moment I swore, for his sake but also for my own, that I would try to see Alex's autism as simply a difference, not a flaw. I started to look into special schools, where a kid like Alex would be seen as normal. In the meantime, I talked to Wendy about getting a specialist in the classroom a couple of times a week, to start helping Alex and the other children understand each other.

At that moment, life changed. I was no longer a writer with a family. I was a mother with a writing habit that I got around to if I had the time. I decided my new career was Alex.

In the process of researching the best schools, diets, and treatments for Alex, I stumbled across something amazing. A short article by Doctor Edward Amend, a child psychologist in Lexington, Kentucky, said there were many cases of highly gifted children being misdiagnosed as having Asperger's. I had never heard of this. It had never been suggested to me by any of Alex's teachers or doctors. Gifted kids, like Asperger's kids, obsess about objects or interests. They are socially awkward. And, if they learned to speak very early, they often confuse pronouns.

I started watching Alex more carefully. Asperger's kids do not have the ability to empathize. But one day, when one of Alex's classmates was crying, I asked him why. "Because she's sad," he told me. I asked him why he thought she was sad, and he said, "Because she wants to go outside." Asperger's kids could not put themselves in another's situation so easily; it would take years of social-skills training for them to do what he had just done.

Asperger's kids also didn't make eye contact, but Alex made direct eye contact when talking to people. He was also a good listener and, it seemed to me, he understood the words he was saying. Beyond that, he even seemed to be deeply philosophical about certain things. One day he declared, "Red is not a color." When I disagreed with him, he said, with a wry grin and direct eye contact, "No, Mommy, 'red' is a word to describe a color."

I remembered that one of the clinicians who evaluated Alex said he wanted to videotape him for teaching purposes, because he had "never seen an autistic kid quite like this."

Maybe that was because Alex wasn't autistic.

Asperger's occurs in about one in every five hundred people. Giftedness of the type that is *confused* with Asperger's? One in thirty thousand. And when it comes to minority kids, even fewer are diagnosed as gifted, because clinicians tended to overlook them. Research done by the National Research Center on the Gifted and Talented indicates that "minority and economically disadvantaged students are not found in (gifted) programs in proportionate numbers." The discrepancy is due to "a variety of historical, philosophical, psychological, theoretical, procedural and political factors," including test bias, clinician bias, and selective referrals from teachers and others.

The teacher who assumed Alex was autistic was not a minority. None of the clinicians who evaluated him were minorities. Alex is a brown-skinned boy named Rodriguez, in New Mexico, the poorest state in the Union. I remembered that when I had suggested to one of the clinicians that perhaps Alex was simply a genius, he had chuckled, as if this could not be possible.

I remembered how, when I was in the fifth grade in New Mexico, a teacher of gifted children came to our class and asked the teacher, in front of the students, to send her smartest pupils to the library to be tested for the gifted program. In a class that was mainly Hispanic, the teacher chose the only three non-Hispanic white boys. Being a fighter, I insisted they take me, too.

I got the highest score in the history of the school.

Within a week of learning about the high rates of gifted children being misdiagnosed as autistic, Patrick and I took Alex to see Maria Teresa Velez, a woman who had grown up with my father in Cuba and was now a professor of psychology at the University of Arizona in Tucson. She offered to let us stay with her for a few days so she could see Alex in action, with the promise that she would tell us honestly what she thought.

After watching us interact with Alex for about two minutes, Maria Teresa laughed out loud and said, "Boy are you two colluding to make this kid the way he is."

For example, she had just given Alex her wallet to play with. When he began to examine the contents, and dropped a credit card on the floor, Patrick jumped to get the card for him. Maria Teresa held up one hand and said, "Daddy, don't do that. Make him get it."

We'd never *made* Alex do anything. In fact, we'd rather let him run our lives, now that we thought about it. And ever since being told he was an Aspie, we'd let him run our lives even more.

Maria Teresa also pointed out that we always referred to him as "we," like when we were going to help him get dressed, we'd say, "Now we're going to put on pants."

"Of course he's confused," she said. "You won't let him separate! This is bad parenting."

Patrick and I were speechless.

Alex looked Maria Teresa in the eye. He answered her questions. When asked how the dog felt to be outside, he said, "Sad."

"There's no way in hell this kid is autistic," Maria Teresa said. "Now give him a bath and put him to bed."

When I tried to stay with Alex on Maria Teresa's bed until he fell asleep, the usual routine, Maria Teresa cut me short. "Let him cry," she said. "God, give the kid some space, Mom. What's wrong with you?"

"What if he falls off the bed?" I asked.

Maria Teresa laughed at me. "First of all, he won't. Second, if he does, so what? He won't die."

The next day, we went to see another specialist, a child psychologist in Tucson with twenty years of experience with autistic kids. She, too, took one look at Alex, who once again made direct eye contact and answered questions, and said, "It is very unlikely this child is autistic or Asperger's. I can't imagine why anyone would tell you he was."

On the flight home from Arizona, we explained to Alex that we would not accept him using the second person to describe himself anymore. Nor would we tolerate the "bad word" obsession game. If he did "bad word" anymore, we said, it would be grounds for a time-out. We had never really given him a time-out for anything.

And—I swear to God—by the time we had dinner in the Gardunos restaurant in the Albuquerque airport, Alex was using the first person to describe himself. Like it was nothing. Like he'd been playing a game all this time. Like he'd been testing limits through language play, which was Maria Teresa's theory. He continued to make mistakes sometimes, and refer to himself as "you." But when we corrected him, he understood. And he changed. Literally overnight.

But I still wasn't convinced. After all, Maria Teresa was a family friend. So, I contacted Dr. Amend, the one who'd written the piece about misdiagnoses that sparked my interest in the first place. We made an appointment for Alex with him in Lexington, Kentucky.

After two days of tests and evaluations, Dr. Amend sat down with us and said that our son was not autistic. He was flabbergasted as to how anyone could come to that conclusion about Alex.

Furthermore, he said, Alex would be "normal" for language and reading comprehension, if he were in first grade. In other words, Alex, at age three, was speaking and reading as well as the average six-year-old. His social skills were delayed, though. But, said the doctor, this was likely due to . . . parenting.

"Not setting enough limits. Overindulgence."

Spoiled? We'd screwed him up?

What?

Patrick and I looked at each other in shock. We'd thought we were the perfect parents. We didn't do any of the bad stuff our parents had done, like spanking and yelling. We didn't let Alex watch TV or eat junk food. He had expensive wooden toys instead of cheap plastic ones.

But what we'd failed to give him, Dr. Amend suggested, was structure and predictability. For two parents who had prided themselves on being attentive and loving, this news was hard to swallow. We hadn't made Alex sleep in his own bed because we thought only fascists did that. We hadn't let Alex cry for anything, because we saw his sadness as our own failure. Our mistake, the doctor told us, was wanting Alex to like us more than we wanted Alex to be prepared for life.

What the doctor was saying felt right, but awful, in the same way it feels when a trusted friend tells you yes, you have gained some weight. It was something we'd suspected but never been able to admit. We weren't perfect parents. And perfect parenting did not mean one big party for Alex. Perfect parenting meant sometimes doing things that made your child throw tantrums, or cry. It meant teaching him how to take care of himself.

Patrick and I talked about why we'd parented this way. The answers had nothing to do with Alex, and everything to do with us. Patrick is a brilliant man, a writer and comic, who'd grown up in a working-class Mexican-American family with very rigid gender roles and traditional ideas. His conservative, anti-intellectual parents had never taken him to a museum. They made fun of him for being a liberal. They had never valued his choices or his personality. They had never made him feel important. So, Patrick had gone out of his way to make sure Alex got to be Alex.

And I? My mother had abandoned me when I was eleven. She was negligent to the point that the state called her an unfit mother. My father was, when I was growing up, a self-absorbed man who routinely forgot my birthdays and lunch money. I had felt worthless to my parents, and I was going out of my way to make sure Alex knew how important I thought he was—to the point that he became frightened, because there were no rules and I seemed to take my cues from him.

What he really needed was a structure to rebel against. Instead, I had followed his lead.

So, we had screwed him up by trying not to screw him up?

Well, not entirely. Alex was also misunderstood. And misdiagnosed. And, Dr. Amend said, it was nothing a little structure and positive discipline wouldn't repair.

Alex, it turned out, was probably the baby genius we'd suspected he'd been all along. But then he got lost in the mediocre school system in Albuquerque, in the hands of teachers and doctors who had likely never come into contact with anyone like him. He was misdiagnosed merely because he was different. I also suspected that, like me, Alex had been dismissed because he was Hispanic. As a high school student in Albuquerque, I was thrown out of honors English because, instead of writing an expository paragraph, I wrote something different, a baroque, curly and unrelated paragraph, which I found beautiful and more interesting. The teacher, who valued conformity, told me I was better off in remedial English. When my father immigrated to New Mexico from Cuba, he was placed in a fourth-grade classroom because he spoke Spanish. He had to sit in tiny chairs, with nine-year-olds who made fun of him and called him retarded simply because his English wasn't perfect.

I had been fighting my whole life against these kinds of reductive, unjust attitudes. But somehow I never thought I would have to start fighting for my son so soon.

When I was pregnant, and dreamed about who my little boy was going to be and what he was going to become, I never thought about autism. And even though finding out he was actually gifted instead was an enormous relief, that diagnosis hasn't been easy, either. People assume that gifted kids excel all the time. The truth is the reverse. Most of them give up. They're too different. They're geeks. Kids make fun of them. They don't learn in traditional ways. They are special-needs kids of a different kind, and most teachers don't invest much energy in children who are smart beyond their years. And so, I'm bracing for a lifetime of fighting. Because parenting, I now realize, is more than just loving your kids and letting their genes "take them where they're going to go." It's loving them and taking the time to figure out who they are. It's loving them and fighting for their right to be themselves. It's doing the research yourself, because ineptitude is rampant. Parenting is fighting, and fighting, and fighting to let this person's soul soar—and never letting anyone, no matter how important they sound, underestimate your child.

Especially yourself.

NEAL POLLACK

One Is Enough

A FEW YEARS AGO, MY WIFE, REGINA, and I were living in a really nice three-bedroom, two-bathroom, one-big-tub-with-massage-jets apartment in Chicago, the kind of place a certain type of couple can rent when they're only worrying about themselves. We didn't make a lot of money, but we acted like we did. We went out five nights a week, minimum, throwing down twenty bucks for some world-music hoo-ha in which we had only marginal interest, eating meals at restaurants that, today, would send our budget into a tailspin. Our only dependents were four cats. Way too many, but cats can deal. We thought nothing of skipping the country for three weeks, twice a year.

One night, as Regina and I relaxed in one of our two enormous living rooms, I took a sip of scotch and said, casually, "Our life is pretty great right now. I don't know if I want to have kids."

Regina's face contorted. "What do you mean?" she said.

"I mean it might be nice to, you know, not have any kids so we can do whatever we want."

A thick rain began to fall outside. Our windows slicked with the torrent. Agonized lightning flashes mixed with organ peals of thunder.

The wind called the lost ships of Lake Michigan home from the deep.

"What do you mean?" she said.

"Um. I don't want to have kids, maybe?"

There's no way for me to write about Regina's reaction without it sounding like dialogue from *The Young and the Restless,* so instead, indulge me while I make it sound like something from Shakespeare, one of the few writers capable of registering the intensity of her response.

"Show your false self no more before my wet-veiled eyes!" she said. "Thine fickle promises are cast upon the rocky shores of hope! O, castle of barren desire! I plead mercy on this forsaken womb!"

"Maybe I do want to have kids," I said.

She sniffled.

"Are you sure?" she said.

"Oh yes," I said. "Very sure."

Regina is adopted. A child, to her, meant an authentic blood connection to the world that she's never had. For years, she struggled with restrictive state laws that barred her from knowing anything at all about her birth parents. After years of trying, she finally tracked them down, but both her birth parents and grandparents wanted nothing to do with her. Regina told me that if we didn't have children, she would spend the rest of her life under a cloud of bitter regret. I could understand that. Besides, I had no interest in living inside a Tennessee Williams play. I liked children, and I was good with them. It's just that one can grow fond of a certain lifestyle.

Regina and I talked about having "children," as in more than one kid. We were childless at the time. But now we have a two-year-old son, Elijah, whose presence, however delightful, has encouraged us to adopt a comprehensive family-planning program. For two years, because of the kid, our lives got lost in a whirlwind of near-hysterical bitterness and compromise. We didn't spend a night apart from Elijah. There was little sex, and the sex we did have wasn't much fun. Vacations suddenly got expensive and, after Elijah started to walk, impossible. Now, at last, the fog is lifting.

Regina has begun to revert back into the woman I married, a

lively, opinionated, almost bossy person who paints every morning, screams at the news, sometimes drinks more wine than she should, instigates hanky-panky in the shower, and likes to put on fancy jewelry at night. But until very recently, she was a mother, and little else.

Elijah was her compass. She pureed his vegetables. Early on, when she figured out he was allergic to dairy, she started monitoring his poop to figure out what kind of milk gave it the best consistency: goat, sheep, or soy? She bought Elijah a UV-protectant two-piece bathing suit from Australia and spent weeks on eBay shopping for his bedroom area rug. I had Elijah two or three hours a day. We called that "Daddy time," and it usually involved going to the park and dancing around to music, with the occasional bath or meal. Otherwise, Elijah was Regina's responsibility.

But Regina was ready to detach when Elijah hit twenty months. She wanted to teach a couple of art classes every week, and she wanted time in the mornings to paint. Daddy time was starting earlier and lasting longer. I often had to stop work when I was in the middle of something to take care of Elijah. And Regina still barely had any time to herself. Every time one of us wanted to do something alone it would take rounds and rounds of negotiations. The conversations went like this:

"You selfish bastard!"

"You castrating bitch!"

"Goddamn it, you're such an asshole!"

And so on. My sex life turned into whatever I could find on the Internet at one a.m. Dinners became sullen affairs eaten in front of reality TV shows. I traveled occasionally for work, but Regina never left Elijah's side.

During this time, Elijah gained what we, as a couple, had lost. In his first twenty months, he was only sick once, for less than twenty-four hours. He got up early, way too early, but he slept regularly, on a reliable schedule. I've never seen a kid eat so many vegetables. We gave him as many nice toys as we could afford, lots of hugs and kisses, delicious animal crackers, a backpacking trip to Colorado done on the cheap, and, most important, time. We read with him, played music for him, and took him to the aquarium store to see the fishies. He got everything we had because there was only one of him.

But Regina and I both agreed that we'd voluntarily sacrificed enough for one lifetime. A few months ago, we finally put him in morning day care. That first day, it was nine a.m. and we had nothing to do.

"You wanna get some coffee?" she said.

"Sure," I said.

At the coffee place, she said, "I don't want to go through it again."

"Thank God," I said.

"I think if I had another kid, I'd start to resent both of them."

"I totally agree."

We both looked around.

"So," I said. "What do you want to do now?"

Newborn babies squall. They hunger. They're helpless blobs of tufty hair and wrinkled flesh, demanding your constant care. Yet they carry benefits. For instance, they're soft. They fall asleep anywhere. We went out constantly during Elijah's first four months. It didn't matter. Restaurants, concerts, even professional basketball games. He slept through them all.

Toddlers throw tantrums. They break things, both by accident and on purpose. They bite and hit and wriggle and climb, and it's impossible to take them anywhere in public for very long except to playgrounds, which are boring for adults. But their benefits are legion. They talk and laugh and sing and play and act a little more like a person every day. They can identify the guitar-playing guy on their bedroom-wall poster as "Johnny Cash."

However, if you combine the two by having a newborn at the same time you have a toddler, your life is utterly ruined. The newborn doesn't let you sleep, and the toddler doesn't let you rest. Suddenly, it becomes a deeply despairing logistical task to leave the house. If we had a newborn right now, our lives would be a total river of shit and piss and puke, instead of the partial one in which we now swim. We have plenty of time to read to and play with Elijah, to take him swimming and to the park. And now that he is a little bit older, and sleeping through the night, Regina and I have some sort of life. When we want to go out, we have a system. After Elijah goes to bed, Regina immediately gets in the car and goes to the 7:30 movie. She comes home, hands me the keys, and I go to the 9:30 showing. Then I come home and we talk about the movie before we go to bed. I still get to go see the bands I want. She goes to her book club and goes out drinking with her friends. With a second kid, our lives would be over.

When we tell people that we're only going to have one child, they're aghast, or at least surprised. How can we commit Elijah to a life without siblings? Because, I say, if he's feeling lonely, he can have a friend over. Won't he be spoiled if he's the only one? Not with our credit card debt. Even now we have to call my parents every time he needs a clothing upgrade. But the cliché I hate most, and the one that gets thrown at us most often, is this: Who's going to take care of you when you're old? This, of course, assumes a fantasy world where adult siblings share equal responsibility for their declining parents. I might as well answer, "The elves will take care of me, of course." Maybe my son will take care of us. Maybe we'll take care of each other, or maybe our second or third spouses will take care of us. Maybe I'll die at fifty. Or maybe I'll have a really sexy nurse whose hot body will keep me going for another decade. Who the hell knows? But don't tell me to have another kid because I may wear diapers myself someday.

I'm just being selfish, you say, or they say, or someone says. Hardly. As everyone predicted, having a kid has *changed my life forever*. Before Elijah, I didn't read Sandra Boynton's *Barnyard Dance* five times a day, I didn't think about whether or not the frozen carrots were organic, I was actually allowed to watch *The Simpsons*, which Regina now claims is "too violent," and dinner wasn't preceded by fifteen minutes of "tickle time." My life is actually better because I have a kid, and my kid's life is better because he's the only one. Some people have large families because of religious beliefs, or poverty, or some combination of unpleasant circumstances. But anyone who deliberately wants lots of kids is either replacing reality with parental martyrdom or planning to run for political office. Either way, your kids are props. Who's more selfish: The parents who have one kid and lavish it with attention, or those who have four and plop them in front of *Toy Story 2* several times a day? I stand up for our choice.

We have a happy kid. So said the mother of a friend of mine, who has three sons, at a barbecue we brought Elijah to last summer. When I think about the joys of having only one child, that barbecue comes to mind. It was at the home of a bachelor, a place that's frequently filthy and always haunted by the piss-smell of the two near-wild mongrel canine-things that live there. No one at the party, except Elijah, was under twenty years old. The yard, a minefield of two years of dog shit, wasn't appropriate for kid's entertaining. I turned up the Dad-o-Meter and went to work.

"Ben," I said to our host. "You got any toys?" He thought about it for a second, and then went to his desk.

"There's this Pokemon character," he said. "You turn him on and he quacks."

"That'll work," I said.

I handed the toy to Elijah.

"What's that?" I said.

"Duh!" said my son.

"Yes. And what sound does a duck make?"

"Dahk Dahk!"

"Right. Ben. You got anything else?"

Within five minutes, I'd procured, for Elijah's amusement, a plastic monkey figure from some obscure anime, a Kelly Osbourne bobblehead, and a Batman comic, which was probably too violent for him, but I just love to watch him run around while he says, "Ba-Man! Ba-Man! Ba-Man!"

I took care of Elijah for a while. Then I had a beer while Regina fed him some salad with rice and olives, a small amount of hamburger meat, and a couple of delicious slices of beefsteak tomato. Elijah rolled around on the couch and made eyes at beautiful women.

Then I heard the telltale shriek.

"That was loud," someone said.

"Time to go," I said.

And that's when the mother of three said, "You have a happy kid." She also said, "Some parents don't know their kids' limits. You guys are obviously great parents."

Well, thank you, ma'am. I think we are, too. But who's to say we'd be great parents if we had two kids, or three, or four? We'd barely be able to function, and I bet our kids wouldn't be as happy, either. With one kid on the string, we enjoyed the party, got to meet some new people, have a couple of drinks, and eat some tasty food. Our kid was a little bit of work, but mostly he just integrated into the crowd and made the afternoon more fun for everyone. With one child, you can have a good time but temper the fun with a little responsibility. A kid also gives you a good out when you want to blow a scene.

When we got home, Elijah was still hungry.

"Well, make him dinner," I said.

"I gave him breakfast and lunch," Regina said. "Why do I have to give him dinner?"

"Because I have to do some weed-whacking in the backyard," I said. "I told you that this morning."

"Can't you feed him dinner and whack the weeds?" she said.

Every day since Elijah was born has contained some variation on this dance. I assume every day will, for the rest of my life. Mostly, I don't complain. Even when I find myself eating at places like California Pizza Kitchen because they have a junior menu and crayons, or wasting Saturday afternoons at Sea World in San Antonio and at the Waco Zoo. But when, for instance, I take Elijah to a big music festival and it's 105 degrees outside and I have to spritz him with a squirt bottle to keep him alive, and he hears Solomon Burke's band start up and says, "Jazz, Daddy, jazz!" I think: My son is so cool.

But he's also a biter. As I write this, he's bit a kid at school three of the last five days. "What'd you do at school today, Elijah?" I ask.

"Bite!" he says.

Great. The kid likes to bite. This is a problem. But we'll solve it, because we always do. At least we've only got one.

AMY REITER

Mama Don't Preach

I JUST GOT OFF THE PHONE WITH A CUSTOMER service representative at my credit card company. I called to dispute a charge—the hospital where I birthed a baby boy seven weeks ago made a billing mistake—and the next thing I knew, I was having an extended conversation about labor and delivery with the stranger on the other end of the line.

She's due in a few months with her second child. She told me the birth story from her first: They overdid it on her drugs and she was totally numb up to her ears for hours. I told her mine: I had a C-section that didn't heal right and had to be (stop reading if you're squeamish) *reopened,* picked at, prodded and allowed to just sit there as a gaping wound for weeks, healing gradually.

Yes, it hurt. More than labor, in fact. I was in the hospital for a week, and a visiting nurse came to tend to me twice a day for more than a month after that, at which time my husband was charged with latex-gloving up and ministering to my slit midsection on a daily basis. Sexy? I think not.

Though my childbirth scenario was a bit too heavy on the pain and hers a bit too light, we both agreed that actually having the baby was completely amazing. True, we'd suffered, but we'd each been rewarded with a big prize—a healthy, delicious child.

"I cried for three weeks afterward every time I looked at her," she shared.

"Yes," I agreed, having dampened many a onesie with tears of both the gently rolling and the sobby, gulping variety. "I've done a lot of that, too. The whole thing is pretty emotionally intense."

Then she said, "I don't understand how anyone can *not* have children. They're missing out on the best thing in life."

At that point, I got off the phone.

Because you know what? Thrilled as I am to be a mother and to hang with this astoundingly adorable little person sprung from within, I refuse to jump on this particular parental bandwagon, the one packed with proselytizers peddling their babycentric life view.

If you're living a child-free life, you probably know what I'm talking about: People who start out celebrating their own decision to have kids and end up casting aspersion on your choice not to—or at least not to have them just yet. They'll carry on—with great concern about your ticking biological clock, of course—about how their lives before children were (and implicitly, your life without them is) empty, lonely, devoid of meaning, even downright selfish.

These procreation proponents stepped up a few years ago, revved up by Sylvia Ann Hewlett's *Creating a Life: Professional Women and the Quest for Children,* in which the author declared that women today are in a "crisis of childlessness." Wait too long, she warns the ladies, and you may suddenly wake up seized with regret and unable to conceive.

I'm past thirty and a brand-new parent. At this point in my life, though not before, having a baby feels completely, euphorically right for me. I'm deliriously happy with my decision to spawn. But who am I—or who is anyone else—to say that having a baby is the best thing for everyone?

Not only do I not think that the only meaningful life is one that includes children, but I'd also go so far as to say that there are people in the world who should not have children.

I'm not talking about people who are, say, prone to abuse. Those people are givens. I'm talking about perfectly nice, everyday folk who don't really want children. People who are happy with their lives as they are and are uninterested in turning them upside down to meet the needs and whims of a growing child.

Like marriage, only more so, having children is an irrational act, a total leap of faith for all who attempt it. If you worked up a cost-benefit analysis of childbearing and -rearing, the cost side would be filled with real sacrifices—financial, physical, emotional—and the benefit side would feature things like "When my baby smiles at me, I go all gooey inside."

Having a baby and raising a child pushes you to your physical and emotional limits . . . and way, way beyond. You need that fire-in-the-belly thing to light your way through the dark patches: the exhaustion from middle-of-the-night feedings, the incessant worry over every random cough and snuffle, the physical strain of hefting an infant who refuses to be put down, the constant suspicion that you and your spouse simply don't know enough to be parents. It's an emotional labyrinth, and you can really get lost in there.

Particularly if you happen to be a woman.

The physical consequences of pregnancy and childbirth alone can be unpleasant and ongoing. I have friends (at least two of them in my immediate circle) who are doomed to lives of maxipad wearing because, since giving birth, they can't sneeze or cough without a little pee leakage.

I have another friend who, more than a year after her son was born, still consistently endures nether-regional pain—mild on a good day, not so mild on a bad one. Another buddy recently told me that since the doctor stitched her up "a little too enthusiastically" after the birth of her first baby, sex has been downright unpleasant.

Not to mention the new sags, bumps, lines, and wrinkles that will keep your body from ever looking the same. "The new normal" is what my husband calls it, but he's just being nice, considerably nicer than I am to myself when I work up the courage to look in the mirror. Factor in the toll that sleepless nights and random bouts of worry—or out-and-out fear—take on your skin, and other odd corporal goings-on, and, well, sister, you ain't no nubile teenager anymore.

And that's just the physical side of things. A friend of mine who just had a textbook vaginal delivery told me she felt so traumatized by the exigencies of labor and delivery that she's planning to start therapy just to come to terms with the emotions it all stirred up. Another friend has been coping with postpartum depression so debilitating she has been unable to return to work as planned.

Even in the best new-baby scenarios—mine, for instance, if you discount the abdominal-wound factor—there are moments of severe self-doubt and self-pity in the midst of the baby bliss. *Is the baby nursing enough? Is he nursing too much? Why won't he go to sleep? Will he ever go to sleep? Is it my fault he won't go to sleep?* You have to summon all your own inner strength—and the help of your partner (if you're lucky enough to have one), family, and friends—to pull through the first few hormonally rocky, sleep-deprived weeks.

Get past them, and you're hardly in the clear. The need to make a living can feel—as a friend who went back to work last week, leaving her fourteen-week-old daughter at home, put it—"like some kind of primal wrong."

Think the trouble's all in her head? Try in her breasts, swollen beyond belief with milk her baby is not around to drink on her normal schedule. Pumping only goes so far when your baby goes on a hunger strike, refusing a bottle and crying incessantly until you get home to feed her from your own body, only to wake you up every two hours all night long because she's starved from her milk-free day. But have fun explaining that to a boss who doesn't understand why you never work past 5:00 anymore or why you're too tired to take on extra work the way you used to.

See you on the mommy track, girlfriend.

And while we're doggedly running around and around it, we can talk about all the things we miss from our old lives. Like going to the movies or the theater or the ballet. Like enjoying a *leisurely* meal at a restaurant. Like getting up in the morning and going to the gym without first negotiating with your spouse for your forty-minute parental leave.

I'm not complaining. I wanted to be here, gazing into my newborn's eyes instead of, say, getting all dressed up and going to the spate of black-tie shindigs I get invited to each spring. The little fella may not say much yet, but he's already a better conversationalist than most of the tablemates I've been compelled to chat with at such events over the years.

But parenthood as panacea? I'm not buying it, and neither should anyone who's not really into the idea of being a mom or pop.

I'm here as a new parent to stand up for all those non-parents out there—the ones who haven't yet made up their minds about kids and the ones who definitely have—and proclaim that there is nothing wrong with not having children. I did it for more than three decades and led what I'd consider a pretty rich life, filled with learning, love, travel, adventure, laughter . . . and other people's children.

You're not being selfish. Your life won't be empty. And you're certainly not destined for a sad, lonely end. People can find meaning in their lives in ways that don't include progeny.

So, the next time some well-intentioned parent harasses you about your decision not to have kids—or at least not to have them yet—just let yourself off society's hook, go out, and live the life you've chosen with no regrets. Find fulfillment by climbing a mountain, jumping out of an airplane, taking a job in Asia or, hell, reading the Sunday paper without interruption. Then tell us breeders about it.

And feel free to gloat.

ASHA BANDELE

My Tribe

ON SEPTEMBER 30, 1995, my husband and I had a honeymoon of sorts, a forty-four-hour jaunt in a trailer at Sullivan Correctional Facility, the New York State prison where Rashid lived. In 1983, at the age of twenty-one, he had been convicted of a gang-related murder that occurred three years before, and on this day, he was in the thirteenth year of a twenty-year sentence. When we met in 1990, I was a twenty-three-year-old college student, teaching and reading poetry to women and men who were incarcerated. I believed then, as I do now, that poems could enlarge a heart, expand a soul. They had done so for me, and I was watching the same thing happen to prisoners and especially to Rashid, who was already a voracious reader, already a man seeking transformation.

Slowly, over the course of a year and after many discussions about changing ourselves and our world, Rashid and I fell in love. With him, there was a width and breadth of dialogue, a level of introspection about the world and about ourselves, that I hadn't experienced with anyone else before. Certainly not with a man. Our conversations—

broad, emotional, unspeakably honest—were, for me, life-saving. Rashid helped me to understand who I was, my relevance in the world, the beauty I brought to it. When I came to him, a failed marriage already behind me, no college degree yet achieved, no direction for my future clearly laid out, I saw my life as a series of mistakes. I saw *myself* as a mistake. He took that lie and set it aside and honored me, my poetry, my dreams—unformed as they were. He asked me to see myself through his eyes until I could love my reflection, treasure it.

Finally, when I could no longer stand not being able to touch this man who, through his insights and warmth, had so touched me, we married in the prison visiting room. An interminable five months later, we qualified for conjugal visits—"trailers," in jailhouse vernacular. The prison issued us a date, and in a small, two-bedroom trailer in a yard on the prison grounds, we made love, again and again. In the bathroom, on the sheets I bought just for the occasion, in the kitchenette while I was making coffee. But more than that, we had a period of semi-normalcy. We made curried chicken and roti, danced slow to Al Green, watched the news, showered together. For five years, this was our life, the life we chose.

I didn't tell Rashid during that first trailer visit that I would never have a baby with him. I knew many women with incarcerated husbands had chosen to do so, and while I did not judge their decisions, I knew it wasn't for me. I didn't want to be a single parent. I couldn't imagine bringing a baby into a prison. I had made the choice to marry someone in jail, but I was an adult, willingly taking on the emotional slop that loving a person who is imprisoned entails. A child is innocent. I could not bring a baby into that world. But I didn't tell Rashid this because at the time, I didn't think I could get pregnant. I'd had so much sex in my life, had twice lived with men, had been married once before, and yet I had never conceived. I went up to the facility and made love to my husband and I did not worry.

Three weeks later I found out I was pregnant.

I never considered keeping the baby. Even as I wept, rubbed my abdomen, claimed my child, listened to Rashid's pleading, I knew I wouldn't keep it. How could I? It wasn't just the prison or my unwillingness to be a single parent. It was also financial. I wasn't working steadily. I was living in a room in someone else's home, barely eking out a living as a writer and poet. I had just returned to school to finally finish my bachelor's degree. What did I have to offer a child, except anxiety and instability? Rashid and I argued bitterly—something we'd never really done before—about the decision, but all he could offer me were religious platitudes about life being sacred. What about *my* life, I'd argue back. Isn't it sacred, too?

Against my husband's wishes, and against the love I felt for the baby who had barely begun to take shape inside of me, I had an abortion on a cold Thursday in November 1995. In an East Side Manhattan clinic that seemed less like a medical facility than a factory, I sat on a hard plastic chair with rows and rows of other women, some far younger than I, some shockingly older. When my name was finally called and I had dispensed with the routine tests, I was led into the ice-cold room where the procedure took place. Despite the two doctors and two nurses, despite all the women waiting just outside for their turn on the table, it was the most alone I've ever felt in my life. As the procedure began, I started crying so hard that the anesthesiologist worried I might not be able to breathe. Still, he placed the oxygen mask over my face and told me to calm down. Fifteen minutes later, I woke up in the recovery area no longer pregnant.

Having the abortion was absolutely the right thing to do at that time in my life. Still, I carried guilt about it for years. Terminating my pregnancy was against the natural order of things. Had Rashid not been in prison and had we been married under normal circumstances, I surely would have had the baby. Or, if he wasn't in prison, and I felt we couldn't have the baby for financial or other reasons, I would have done much more soul-searching. The choice would not have seemed so obvious. But that day, lying on that examination table, I knew that no matter what difficulty, I could never have another abortion. The weight of destroying something that was created from a place of great love was unbearable. I did it once. To do it a second time would literally crush me.

In July of 1999, I was in California, in the midst of a book tour, when Rashid and I were issued a date for a trailer. I flew home, packed everything we needed—food, sheets, towels, birth control. My period was due in about a week. We made love and celebrated the book, then made love some more. We talked about our dreams for the future, about what was next for me on the literary horizon, about how perfect things would be when he came home.

A week later, when my period was a couple of days overdue, it didn't even occur to me that I might be pregnant. But after six or seven days, I decided, reluctantly, to buy a test. Fast as I could pee on a stick, dark pink double lines appeared. When Rashid called the next day, I told him directly: "We're having a baby."

Unlike our first pregnancy, I never thought *not* to have this baby. That she could create herself in spite of contraceptives and at a low-ovulation moment indicated to me that her presence was predetermined. Also, I was getting older; I was thirty-two years old when my daughter was conceived. I might not have another chance to be the one thing I knew I always wanted to be: a mother.

Still, there were fears. I was a freelance writer when I became pregnant, and I worried about finding a stable job and decent child care. I was terrified of going through labor alone, and surviving the sleeplessness of the baby's first months. And although I knew I couldn't control it, I worried about what people would think. Did people look at me—a seemingly single black woman—and see a statistic? The idea of that was intolerable. I tried to fight back, but my attempts were pathetic.

With the aid of lotion, water, and soap, I squeezed my wedding rings onto my fingers long after they'd swollen into fat little sausages. In doctors' offices and later at the midwifery center where I gave birth, I looked at no one, sat up proudly, made calls on my cell phone to girlfriends, figuring Rashid, or "my husband" into every conversation. And when the time came for birthing classes, I paid for private ones, so I would not have to sit with other women as their partners caressed their bellies. I didn't want to risk judgment, or the rude questions no one dared ask: who's the father and where is he anyway?

What I worried about most, though, was how I would protect my daughter once she was born. Would I be able to raise a black girl safely in a world that seems only to expand in its ability to hate and destroy? In a culture tipped toward death, with blacks and women often the stand-ins for the bull's eye, would the life of my girl be honored by anyone other than me? The rates of drug and alcohol abuse among young people, the sexual and physical violence glorified in pop culture as though it were sexy, the girls who at eight years old are giving blow jobs in school stairwells, and the groundswell of so-called good girls out on the stroll in Brooklyn, Atlanta, even the suburbs, scared the shit out of me.

Not because I judged or disliked any of these kids, but because I'd been my own version of them. Against a Manhattan backdrop, my sister and I were given music and dance lessons, horseback riding, swimming and art classes. I attended fine private schools, and went to the ballet. I saw Judith Jamison dance "Cry." But it was also a childhood punctured by loneliness and violence. What my parents gave me in terms of education and culture was no armor against the excesses and dangers of high school, of growing up in an urban environment too large, too unwieldy to notice its children coming undone. I began high school at twelve, graduated at fifteen, and while I could put on makeup and heels and stumble behind the older girls, I could not negotiate the social settings—the bars and clubs—where I found myself. There were date rapes by men much older than I. And there was always racism. "Piss-colored nigger-rican" is the term I remember best. When my child reached an age when she, too, would have to negotiate the streets and sex, would she know how to speak to me, the way I was unable to speak to my parents? Would I know how to listen to her, hear what she was telling me, read between her lines?

I had both of my parents present, loving me. My daughter would not only just have me, she would also spend at least the first three years of her life passing through prison metal detectors. And it would only be three years if Rashid made parole his first time before the board, which was unlikely. No matter what people said, how they tried to reassure me that my daughter would not remember the steel bars and scowling guards, I was sure her tiny developing brain and heart and body would be impacted by those things. These concerns, more than the physical challenges of pregnancy, kept me awake through the nights, wondering and worrying: Where could I go, where could I live and raise my child safely? I wanted to run, go live off the grid, have my child, tell no one, keep her forever in my womb.

But disappearing was not an option. I had to stay connected to Rashid, even as my trips to the prison became more and more hellish. The prison guards, normally disdainful of family members, became especially awful once I swelled with life. During one visit, I requested toilet paper four times so that I could please, *please* go to the bathroom, but they just ignored me. Whatever disrespect I could accept for myself, I surely wouldn't be able to tolerate toward my child. The love between Rashid and me could no longer disguise the blight, the stench of the prison, and a terrible question took form in my head and my heart. Once the baby was born, would I still be able to navigate the prison and maintain my relationship with my daughter's father, the person who was closest to me in the world?

But the rounder my tummy became, the more those fears either fell away or shrank to a manageable size. I gave myself over to faith. Eventually, I found work and child care, too. And I made it through labor—quite easily, actually. My water broke at exactly 12:01 a.m. on April 14, 2000. Less than eleven hours later, my daughter, Nisa, was in my arms, and latched to my breast. Seven hours after that, we were home, entertaining people and eating gourmet takeout pizza. Fifteen days after that, on a clear, bright Saturday morning, we were up at the prison, and I was placing Nisa's tiny body in her father's arms for the first time. He cried at the sight of her, and I cried too, though not from amazement or joy or some sort of familial connection. I wept from the frustration of not having a place in that visiting room where I could breast-feed my baby, a place in that prison where we could be, at even the most basic level, a family. I wept because even then, even as Rashid and I held our child together for the first time, I knew what would come.

"You're such a good mother," my sister or a close friend would say, before adding, "How do you do it?" I heard this a lot during the first two years of Nisa's life, when I was writing a novel, working full-time, and meeting my deadlines. I told people that it was hard, impossibly hard to be a single mother. I hated having to be the sole emotional and financial provider for my child. The pressure was too great. If you slipped, there was no one there to catch you. But worse, there was no one there to catch your baby. Somewhere between potty training, play dates, speaking engagements, bylines, and balancing the household books, I've lost pieces of myself I am only now trying to reclaim. I started smoking cigarettes again. I spent too many nights on the couch after Nisa went to sleep, drinking wine and crying. There were times I was sure I had lost my mind completely, where loneliness and hurt and anger and frustration threatened to define the whole of me. And all of this made it impossible for me to keep my marriage together.

Caring for Rashid and Nisa at the same time turned out to be more than I could bear—the weekly treks through metal detectors, the parts of my spirit that always seemed to get snagged by the razor wire that was everywhere. I took Nisa to see her father every month, but for me to be romantically entangled with Rashid when I most felt his absence became too painful. When it was me alone, the hardships that accompanied prison marriage were bad enough. With my daughter part of the equation, it became excruciating. Every time I saw her grow, change, fall down, stand up again, I was reminded that the only other person who loved her as I did was not there to bear witness. I had no one with whom to share the everyday beauty and wonder of my child. No one who would ever lose an hour, as I still regularly do, just watching her sleep. How could I live with that?

"I can't play house or marriage anymore," is what I think I finally said to Rashid who sat there, refusing to fight me, refusing to give me one more battle to wage. "I love you," he said quietly. "I love you too," I said. "But right now, I need the real thing or I need to woman up and do this on my own."

Despite all the sorrow, the breakup, the letdowns, the entire days I suspect I might have fallen over the cliff into mental illness, in between those moments and the deadlines, Nisa and I travel, in New

York City and across the country. We have marveled together at the differences each place offered: the wide deserts just below the snow-capped mountains of southern California and the redwoods in the north; the swaying palms and rainbow fish of Sanibel Island; the alligators and swamplands of South Carolina and the hot, wet greener than green of Mississippi.

We lose ourselves in all that color and life everywhere, including in the parks and gardens of our urban landscape. We make intricate plans for the trips we have not yet taken together to Paris, the Eastern Cape of South Africa, and Bora Bora. Nisa wants to climb the volcano in Costa Rica which I scaled three months before I became pregnant with her. I promise her we will do that trip, and also one to Baja some January, during the migration of the gray whales.

We embrace it, this life, bathe ourselves in it, retain the memory in stories we whisper to each other when it's late and dark but we want, still, to hang on. "That's how we do it," I tell my sister or anyone else who asks. And it's true. But adventures weren't what I had in mind when I decided to be a mother.

There was the promise I'd made to Rashid and to myself. And there was the creeping age thing. But really, more than anything else in the whole world of reasons I could pick from about why I knew I had to have Nisa, it all comes down to one simple fact: I wanted someone who looked like me. This need was an unconscious one, something I didn't understand before she was born, when all the logistics of childbirth and rearing took center stage. But now, fully ensconced in motherhood, I can slowly, carefully begin to examine something I have spent a lifetime both running from and into. Having been adopted near my third birthday, I had never seen, at least to my memory, one person who looked like me, or shared my particular genetic makeup. So there it is, perhaps the whole of the story, certainly the heart of it, buried down here, closer to the foot of it. Of necessity, I've had to have little faith in blood ties. But I longed for it, longed for what I imagined it carried.

Somewhere in me has lived the story of a child who was unwanted. That changed, sort of, when the social worker and my parents—the only ones I've ever known, the only ones who have ever taken responsibility for me—scooped me up late one winter night out of a foster home rife with trouble that no one, even now, will fully disclose to me. What I tell myself to this day is that my parents only came to my rescue because they'd already met me once or twice before. It was apparent that there was something wrong in the foster home I was rescued from, so my social worker would periodically take me to meet potential parents. It is said that my mother, father, and I got along famously from the beginning. As their version of the family history goes, I was a very pretty child, startlingly verbal, inquisitive and affectionate, an all-around charmer.

But what if I hadn't been? What if I had been obviously flawed? Would my parents have chosen me? What qualities must a child possess to be chosen, accepted, taken in, and loved? And not just children who are adopted, but all of our children? Is there some predetermined formula that makes one kid count and another counted out? What if I had been shy or a crier, not too bright, maybe a little funny-looking? Would I have been relegated to multiple foster or group homes?

I know if I tell the truth, the whole truth, then I have to confess that this is the story behind the story I tell the world about who I am. My story is not about a charming black girl who graduated high school early and can discuss opera or hip-hop, depending on my audience. My story is about a girl who believes that if she is not perfect, she will be left behind. It's the ultimate childhood horror about being the last one picked—or not getting picked at all—when the kids on the playground are choosing sides. Here's the point. I wanted someone who would love me, flaws and all, because, well, they *had* to. We would be family, after all. We had the damn DNA to prove it.

Unlike friends of mine who had huge, almost unmanageable biological families and a genealogy that could be traced down through generations—a great-great-grandmother's face that comes back to visit in the eyes, smile, shape of progeny years later—I had no such beginnings. If I wanted it, and, to be sure, I did want it, I had, then, to create it. My tribe, my history, the one I chose begins with Nisa. She is my giggling, demanding, fearless four-year-old. She is my family tree, branches, leaves, and all.

Most mornings, I am awakened by my daughter's laughter and messy kisses planted on my cheek, in my hair. Her eyes ablaze with mischief, wonder, excitement, and hope, Nisa's query to me each sunrise is the same: "What's our big adventure today, Mommy?" I grin back at my beloved, my child, and my mind begins to work. But before I come up with a plan, this is what I think each time she asks: Yes, Beloved. Our big adventure, indeed. *Ours.*

ANDREW LEONARD

Road Trip

Dublin, California, headed east on I-580 to I-5, November 2004

My kids are discussing number theory in the back of the minivan. Seven-year-old Eli is explaining how multiplication works to ten-year-old Tiana. Tiana's contribution is a short discourse on the role of the number zero, a digit whose awesome powers Eli considers "crazy."

This is going to be a good ride, I think. From the sound of their chatter, they are relaxed and ready to road-trip. We have just begun a six-hour journey from Berkeley to my grandmother's house in Lakewood, a suburban city just south of Los Angeles. We are old hands at this jaunt—we know every rest stop, gas station, and highway interchange along Interstate 5's spear-thrust through California's vast Central Valley.

Our rituals are all in place. The Beatles' *White Album* is playing on the stereo, because all road trips must begin with the sound of the jetliner that opens "Back in the U.S.S.R." We picked up some chicken wings and raspberries from the grocery store, and the food is carefully balanced on top of a small cooler wedged between their bucket seats, right in front of a garbage bag. They are cocooned amid a swathe of blankets, pillows, and favorite stuffed animals: "Liony" for Tiana, "Alligatey" for Eli.

The sun is beginning to set, because long experience has taught us that when you're trying to make good time on the Berkeley-Lakewood express, you start at twilight. The kids eat their dinner, and then they drift off to sleep, lulled by the familiar rhythm of the minivan's passage. And I cruise, my world comfortably and completely reduced to the contents of the car.

I never expected, before becoming a parent, that some of my favorite moments of fatherhood would arrive while driving seventy-five miles per hour between the stockyards of Coalinga and the Tejon Pass. Looking back, maybe it isn't so surprising that that is where I grasped some of the most important lessons on how to be a good dad. When you're on the move, you define who you are and what your relationship is with your cotravelers at every step. You learn fast.

But what I find really intriguing is not how I found out how to be a father on the road, but how my kids learned how to be my children.

Ascending the Altamont Pass between Livermore and Tracy, I-580, spring 1994

The baby is sleeping. We hit the road just before naptime, and we're going to see how far we can get before she wakes up and starts demanding attention.

We are outfitted for a major land-war. We have car seat, stroller, and portable crib. We have a diaper bag, canisters of baby powder, a vast assortment of rattles, and a hefty supply of cloth diapers. (Our friends think we're crazy for taking cloth diapers on the road, but hey, we're from Berkeley.) In the rash exuberance of new parenthood, we think nothing of crossing the great state of California with an infant in tow. Sure, we won't make the same time we did when we were unencumbered—but match us up against anybody *else* with a six-month-old, and we're confident our record will be impressive.

It's the pre-minivan era, and we're driving a behemoth 1967 cherry red Ford Galaxie convertible, which is not exactly ideal for total baby comfort. The radio is broken, and the engine leaks oil, and you've got to check the water in the radiator constantly.

But we have style. We're feeling good. We're finally back on the road after several sleepless months at home, holed up with a newborn. Jeni and I have always loved to travel. We've backpacked in Southeast Asia, road-tripped across the country, from Berkeley all the way to northern Florida, sought out the obscurest, most roundabout routes, and driven nonstop through the night in headlong marathons. Now, for the first time, we're finally heading out on a serious car trip as a family. We've got the provisions, we've got P. J. Harvey on the boom box. So far so good.

Jeni and I plunged into parenthood in much the same way we hurtled into marriage. It seemed like a good idea at the time, but we didn't give it a whole lot of thought. Kind of like hitting the road without knowing exactly where you are headed, but on the general assumption that it's time to roll. We knew we wanted to do it, but I don't recall extended consideration of *why*. One day, we said to each other, hey, isn't it time to start a family? The next day, Jeni went off the Pill, and within a couple of weeks, she was pregnant. Somewhere, a switch got flipped, and we went from being carefree lovers to prospective parents.

The baby is sleeping.

I-5, somewhere south of Fresno, summer 1995

It's hot. The kind of baking Central Valley heat that is great for ripening peaches and almonds but is hell on a one-and-a-half-year-old baby who is sick of being in her car seat, suffering from mild diarrhea and utterly unwilling to nap for one more second. The gas station in this benighted section of near desert has no changing table, of course, so an impromptu changing station has been set up in the back of the new, but already slightly battered, minivan.

We no longer bother with cloth diapers on the road—it's all about convenience now—but changing the baby is still a chore. And it's my turn, as well proven by the glare Jeni gives me when I not-so-innocently ask who dealt with the last nasty diaper.

I am only slightly less grumpy than my daughter. We are making terrible time. Tiana only slept for half an hour before starting to whimper, and since then it seems as if we've stopped at every exit. She's teething, and we forgot a teething ring, so we tried to find one on the way. At one point, she was howling so loud that we were forced to stop in the middle of absolutely nowhere, and wait, impatiently, while she played in the dirt on the side of the road.

I'm going to lose my mind if I hear Raffi's version of "Baby Beluga" one more time. Neither Jeni nor I have had much sleep in months, and our interaction with each other has been reduced to a series of militarily terse interchanges that deal mostly with logistics. *Where did you put the diaper-rash ointment? Stop the car. I have to breast-feed.*

Just one and a half years into the age of parenthood, and road trips are no longer fun. They're tedious. Like so much of parenting a small child, they are drudgery, something to be gotten through rather than savored. The first few trips with the baby were a novelty—a learning experience: this is how it works now. But I'm not at all sure I like what I've learned—how every coo of happiness from the baby is matched by a howl and a whimper, how much work it takes just to keep it all together. I long for the days when I zipped through the Valley, making just one stop at a gas station, wasting not one second. But my baby will not comply. Yes, yes, I remember all those warnings that after my child was born I would no longer be the master and arbiter of my own destiny. But deep down, I never really *believed* it. I thought I could outwit my kids, that my will would persevere over theirs. I miscalculated.

I'm ready for this road trip to be over. Two hundred miles from L.A. But it feels like two thousand, and I know that by the time I get there, I'm going to be too exhausted to enjoy it. And then it will be time to turn around and come back.

Just south of the Harris Ranch, headed north on I-5, fall 1999

It is the Sunday after Thanksgiving, and we have made a disastrous strategic error. In past years, we left Los Angeles at sundown, counting on the kids (now there are two) to sleep while we burned rubber in the night. But arriving home in the wee hours and then starting the work week, short of sleep and with cranky kids, presents challenges that, as time goes by, we find increasingly wearying. Besides, there's a friend we want to visit in West Hollywood, so we decide to bite the bullet and return home after seeing her, in the middle of the day.

It's a nightmare. Half of California is on the move. I've seen slow traffic on this highway before, but nothing like this. There are stretches of road that go nearly thirty miles between gas stations, and still the traffic is stop and go, bumper to bumper, crawling along at ten to fifteen miles per hour. Time has no meaning in this wasteland. The land is empty, except for the occasional orchard. It is a vast, unpopulated expanse, save for these strips of highway, jammed with traffic sludge.

Even worse, the Central Valley's notorious tule fog has settled in, a thick blanket of cotton that reduces visibility to just a few yards. So even when the traffic, on rare occasions, does speed up, there's always the danger of a sudden slowdown, announced only by a car's back end suddenly emerging from the fog, dead still.

My marriage isn't doing much better than the traffic, though I don't know it yet. It's all of a piece: the parenting, the travel, the stresses of balancing work and relationship and children. It comes out later in couples therapy—Jeni resented the grueling airplane trips back to the East Coast to visit relatives every summer and Christmas. And rather than making the same drive to Lakewood every six months, she'd have preferred exploring some new territory, taking a less preplanned route. Who, in their right mind, she wondered, would condemn themselves to I-5 in the Valley, again and again and again?

It's quiet inside the minivan, and what conversation there is is directed at the mind-numbingly bored children.

In my bleakest moments, I can see this traffic jam as a metaphor. Once you become a parent, your actions become greatly circumscribed.

You can't just hit the road without planning. You're stuck. Kids are an incredible time-suck—keeping them fed, clothed, healthy. Short of sleep and irritable, you start to feel that you're running in place, and you wonder, is this why I became a parent?

But you don't actually have too much time to philosophize. Somewhere, a diaper needs changing.

On this trip, as mind-numbing as it is, there is actually one sign of hope. The kids are not so bored that they are making our lives any worse. On the contrary, to my amazement, they are doing pretty well. They've got plenty of food, they're busy with some books of stickers we got at the last gas station, they are amusing themselves. Tiana's musical taste has dramatically evolved. She is five years old and likes Tom Waits, Madonna, and No Doubt. She doesn't mind Sonic Youth or Nirvana. She even grooves on P. J. Harvey. Raffi is history.

The kids are gutting it out, making the best of a bad situation. I feel guilty for subjecting them to this hell, but respect their good spirits. They've been on enough road trips that they can deal with one that goes awry.

Pismo Beach, Highway 101, summer 2001

I'm soloing with the kids now, working my way up the coast on Highways 1 and 101. I've learned something. It's a July fourth weekend, and I took an extra day off of work, so we could have a leisurely ride back home.

I've been telling them for weeks that this would be the plan. We can stop at every beach along the way, if they want. We'll keep our options open, take off for a hike if the countryside looks good.

Better late than never, I guess. The marriage has dissolved, and parenting, 50 percent of the time, has taught me that a straight line is not the shortest distance between two points. It's sometimes easier—on the kids, on me—to take a leisurely, we'll-get-there-when-we-get-there approach to life.

Out there on Highway 1, as I appreciate the stunning panoramas of Pacific Ocean clashing with California cliff, and listening to the oohs of my kids at each new explosion of spray hundreds of feet below, I sometimes wonder, would I still be happily married if I'd learned this lesson earlier? If one day, stuck in a traffic jam, I'd suddenly said to Jeni—hell, forget about going to Lakewood, let's just stay here, get a motel room, and hang out at the beach? Why are we killing ourselves?

Musings like this are pointless, of course—there's no going back and taking that detour now. There's also, paradoxically, a certain freedom that comes with solo parenting. My mom, who divorced my father when I was twelve, recently told me of the unexpected joy of being on one's own—you don't have to ask anyone's permission to do anything, she told me. You just decide.

I see her point, but it's not completely accurate. I still do need to ask permission—of my kids. It's hard to say whether Jeni and I could have fixed things—marriages have their own cycles of ascent and decline—not every one is fixable or should be fixed. But my relationship with my kids is a different matter. I'm in this with them for the long haul.

And they have wrangled me into cooperation in a way their mother never managed. They've made it clear that their pleasure is my pleasure. That it's a lot more fun not to try to make the best time, that an hour building sand castles is worth ten on the interstate.

But I've also wrangled them.

After two beaches, a cool diner for lunch, and some spectacular Highway 1 scenery, they are both beginning to drag. We had considered a stop at the Monterey Bay Aquarium or a hike around Big Sur, but I'm wondering if that makes sense. Still, the decision is up to them. I ask: What do you want to do?

We want to get home.

All right. Daddy can do that. Strap yourselves in, kids.

Just south of Highway 140, headed north on I-5, spring 2003

"There it is! In-N-Out!"

Jubilation reigns inside the minivan. My kids have not been shy to declare their level of starvation for the last hour, but they have nixed McDonald's, Burger King, KFC, Taco Bell, and everything else we've seen since crossing the Grapevine into the Valley.

This is mainly Tiana's doing. She has decided to renounce all fast food except In-N-Out. Why In-N-Out, she is often asked? Well, she answers, her face getting serious, did you know that In-N-Out is one of the few fast-food chains that provide health care benefits to all its workers? Did you know that its French fries are made fresh?

I can only smile when I hear her lecture. This is not my doing. I never thought to impose any anti–fast food ideology on my kids. I suspect that their mother has been indoctrinating them, but ultimately, this is all about Tiana, the crusader for social justice. By age eight, my little girl has already marched against wars, protested at the state capitol against teacher layoffs, been interviewed on KPFA, Berkeley's left-wing radio station.

On this trip, she has decided that McDonald's is evil and she will never cross its threshold again. Eli, who has complete faith in Tiana on such matters, follows right along.

And so will I, despite my lifelong love of evil McDonald's fries. But if my daughter says we must no longer patronize their business, I will agree. I am, in fact, gleeful. When you start raising your children, you often think about how you want them to turn out, and how you will attempt to mold their progress. But you learn pretty quickly that you have less control over what happens than you might have hoped.

And so I scrutinize the maps and scroll through the Web pages, searching for In-N-Outs and planning our route. My obsessive compulsion to plan, and, when necessary, make good speed, plugs in nicely. I can tell the kids—no, let's not stop here, there's an In-N-Out an hour away. And they're down with that.

There is only one In-N-Out along I-5 in the Central Valley, as far as I can tell, almost exactly halfway between L.A. and Berkeley. It becomes our midway Mecca—a stopping point to look forward to, a place to rest and recharge. Yes, it's another godawful highway interchange that but for the grace of the interstate would be uninhabited by man or beast, but to us, it's an oasis. It's *our* In-N-Out. We eat our burgers and fries while sitting at ugly plastic tables barely shaded from the hot sun, pestered by flies and assaulted by the sounds and smells of a thousand passing cars. And we love it.

We are getting pretty good at this road-trip thing.

I-5, headed north, about two hours from Berkeley, fall 2004

We've broken with tradition. Tiana wants to hear the *White Album* in the middle of the trip as well as the beginning. Of late, she's become obsessive about lyrics, and there's a lot to chew on here. Why don't we do *what* in the road, Daddy? Then later: Daddy, you know why I like the Beatles? They don't just sing about love.

These are words, that when heard from your child, break you into tiny little pieces. I guess she didn't have any choice about being a Beatles fan, since I was her father. But then, I didn't have any choice, either, because my parents force-fed me each new Beatles album as it came out. Parents, even as they are trained by their children in the proper way to do diapers, and drive long distances, train their children at the same time, to drive long distances without whimpering, and discourse on the hidden meanings of Beatles tunes.

We are returning from an epic road trip that included beaches, amusement parks, movies, relatives, and random adventures. We had no time for a leisurely trip back up the coast—that would have required hours that simply aren't available. But I'm not worried about asking them to grin and bear it. We have full faith in each other at this point. They know that when they need a break from the trail, an exploration into the unknown—it'll happen. And I know that when it's time to just drive, they'll sit back, discuss number theory, ponder rock lyrics, or explain to me the evils of the world. So it's back on the Five.

We pass by the stockyards of Coalinga. We've been here a hundred times before, but somehow, it seems, the kids have never seen it at a moment in broad daylight when the cattle are pushed together in tight masses as far as you can see. It's quite a sight, those placid bovine millions, and it's one I have always appreciated, since it seems to lay bare the reality of our fast food nation. Around the next bend, I warn the kids, you are about to see something that is a key part of the world of McDonald's, and yes, even In-N-Out. As we crest a hill and the vista spreads before us, they both break out in cries of wonder.

Experiencing that wonder through them is why, I know now, I became a parent. I may not have known it consciously before they arrived, but it seems to me that that is what pushed me here, and what keeps me going. Witnessing Tiana rage against injustice, I feel strengthened in my own politics. Watching Eli's eyes widen at the sight of a million cows, mine crack open a little further. I settle comfortably at the wheel, knowing that they are secure behind me. It was a lot of work raising these kids, and having them raise me, but now that we've figured each other out, we make a pretty good working team.

The road beckons.

CONTRIBUTORS

LAURIE ABRAHAM is *Elle*'s editor-at-large. Her work has appeared in *New York* magazine, the *New York Times Magazine*, the *New Republic, Salon*, the *Chicago Tribune,* and *Glamour*. She contributed an essay to the bestselling anthology *The Bitch in the House: 28 Women Tell the Truth About Sex, Solitude, Work, Motherhood, and Marriage* and is the author of *Mama Might Be Better Off Dead: The Failure of Health Care in Urban America*, which was named one of the ten best books of 1993 by *USA Today*. She lives in Brooklyn with her husband and two daughters.

ASHA BANDELE is the author of two volumes of poetry, *Absence in the Palms of My Hands* and *The Subtle Art of Breathing*; an award-winning memoir *The Prisoner's Wife*; and a novel, *Daughter*. She is currently writing a memoir about motherhood and the prison system. She lives in Brooklyn with her daughter.

LOUIS BAYARD is the author of the novel *Mr. Timothy,* a *New York Times* Notable Book and one of *People* magazine's ten best books for 2003. His next book, *The Pale Blue Eye*, will be published by Harper-Collins in March 2006. His previous books include *Fool's Errand* and *Endangered Species*. His reviews and articles have appeared in the

Washington Post, the *New York Times*, *Ms.*, *Salon*, and *Nerve*. He is also a contributor to the anthologies *101 Damnations* and *The Worst Noel*.

AMY BENFER has worked as an editor at *Salon*, *Legal Affairs*, and *Paper* magazines. Her work has appeared in *San Francisco* magazine, the *San Francisco Chronicle Book Review*, the *New York Times Book Review*, *Glamour*, and the *Believer*. She lives in Brooklyn with her daughter.

ELINOR BURKETT, a former reporter for the *Miami Herald*, is the author of eight volumes of non-fiction, including *Baby Boon: How Family-Friendly America Cheats the Childless.* Her writing has appeared in the *New York Times Magazine*, the *Atlantic Monthly*, *Elle*, *Mirabella*, *Harper's Bazaar*, *More*, *Metropolitan Home*, and *Salon*, among others.

MAUD CASEY is the author of the novel *The Shape of Things to Come*, a *New York Times* Notable Book of the Year for 2001, and *Drastic*, a story collection. Her second novel, *Genealogy*, will be published in 2006. She teaches in the MFA program at the University of Maryland.

LAKSHMI CHAUDHRY writes a column for AlterNet.org. Previously a staff writer at *Wired News*, she has written for *Mother Jones*, the *Village Voice*, *Bitch*, and *Ms.* She is the coauthor of *Start Making Sense*.

LESLEY DORMEN's fiction has been published in the *Atlantic Monthly*, *Open City*, *Five Points*, and *Glimmer Train*, nominated for Best American Short Stories, and included in the anthology *20 Over 40*. Her essay, "Planet No" appeared in *Unholy Ghost: Writers on Depression*. She has been a longtime contributor to *O: The Oprah Magazine*, *Elle*, *Glamour*, *Ladies Home Journal*, and other magazines. Her collection of fiction *The Old Economy Husband and Other Stories* will be published in 2006. She teaches fiction writing at The Writers Studio in Manhattan, where she lives with her husband.

MICHELLE GOLDBERG, a staff writer at *Salon*, is the author of *Kingdom Coming: The Rise of Christian Nationalism*, to be published in June 2006 by W. W. Norton.

JOAN GOULD is the author of *Spinning Straw into Gold: What Fairy Tales Reveal About the Transformations in a Woman's Life* and *Spirals*, a family memoir. A former "Hers" columnist for the *New York Times* and the author of many articles for that publication and others, she is also an avid small boat sailor. She lives in Rye, New York.

STEPHANIE GRANT's first novel, *The Passion of Alice*, was published in 1995 and was nominated for Britain's Orange Prize for Women Writers and the Lambda Award for Best Lesbian Fiction. Her work has received numerous awards, including a fellowship from the National Endowment for the Arts, an Ohio Arts Council Grant, and a Rona Jaffe Award. She currently teaches fiction and non-fiction at Mount Holyoke College in Massachusetts. Her second novel, *The Map of Ireland*, is in progress.

KATHRYN HARRISON is the author of the novels *Envy*, *The Seal Wife*, *The Binding Chair*, *Poison*, *Exposure*, and *Thicker Than Water*. Her non-fiction includes the memoirs *The Kiss*, *Seeking Rapture*, *The Road to Santiago*, and *The Mother Knot*, as well as a biography, *Saint Thérèse of Lisieux*. She lives in New York with her husband, the novelist Colin Harrison, and their three children.

MAGGIE JONES has written for the *New York Times Magazine*, the *Washington Post*, *Mother Jones*, *Salon*, and *Elle*, among other publications. A former correspondent for the *Philadelphia Inquirer*, she has received fellowships from the International Reporting Project, the Japan Society, and the Sundance Institute. She lives in Newton, Massachusetts.

ANNE LAMOTT, a longtime columnist for *Salon*, is the author of six novels and four works of non-fiction. Her latest book, *Plan B: Further Thoughts on Faith* was a *New York Times* bestseller.

LORI LEIBOVICH is a senior editor at *Salon*. Her work has appeared in the *New York Times*, the *New York Observer*, the *Washington Post*, *Elle*, *Harper's Bazaar*, and in the anthologies *Mothers Who Think* and *The Real Las Vegas*. She is the founder and editor of the Web site Indiebride.com. She lives in Brooklyn with her husband and son.

ANDREW LEONARD is a staff writer at *Salon*. He is the author of *Bots: The Origin of New Species,* which the *New York Times* called "A playful social history of the Internet." He lives in Berkeley with his two children.

JOE LOYA is an essayist, playwright, and contributing editor at the Pacific News Service. His opinion pieces have appeared in the *Los Angeles Times*, *Newsday*, the *Washington Post*, and other newspapers. In 2000, he was the recipient of a Sundance Writing Fellowship and a Sun Valley Writer's Conference Fellowship, and in 2005, he received a Soros Fellowship. He is the author of the memoir *The Man Who Outgrew His Prison Cell*.

RICK MOODY's works include *The Ice Storm*, *Purple America*, *Demonology*, and *The Black Veil*. His most recent book is a novel, *The Diviners*.

PETER NICHOLS is the author of four books: *Sea Change: Alone Across the Atlantic in a Wooden Boat*, *Voyage to the North Star*, and the bestsellers *A Voyage for Madmen* and *Evolution's Captain*. His journalism has appeared in the *New York Times*, *GQ*, *Outside*, and *Salon*. He has taught creative writing at Georgetown University, Bowdoin College, and New York University in Paris. Before turning to writing full-time, he spent ten years working as a professional yacht captain, living and cruising aboard his own wooden sailboat. He is currently at work on a travel book to be published by HarperCollins in 2006. He lives in Maine.

NEAL POLLACK is the author of *The Neal Pollack Anthology of*

American Literature and the rock-'n'-roll novel *Never Mind the Pollacks*. He contributes regularly to *Vanity Fair* and writes the monthly "Bad Sex" column for *Nerve*. His parenting memoir, *Daddy Was a Sinner*, will be published by Pantheon in 2006. He lives in Austin, Texas, with his wife and son.

AMY REITER is a writer and editor at *Salon*. Her work has appeared in the *New York Times*, the *Washington Post*, and numerous other print and online publications. She lives in Brooklyn with her husband and two children.

AMY RICHARDS led the Third Wave Foundation for ten years and is the coauthor of *Manifesta: Young Women, Feminism and the Future* and *Grassroots: A Field Guide to Feminist Activism*. She is currently at work on her third book, *Opting In: The Case for Motherhood and Feminism*. Her writing has appeared in *The Nation*, the *Los Angeles Times*, *Bust*, *Ms.*, and numerous anthologies, including *Listen Up*, *Body Outlaws*, and *Catching a Wave*.

DANI SHAPIRO is the author of four novels: *Playing With Fire, Fugitive Blue, Picturing the Wreck*, and *Family History*. She is also the author of a bestselling memoir, *Slow Motion*. Her fiction, nonfiction, and reviews have been published in the *New Yorker*, *Granta*, the *New York Times Magazine, Elle, O: The Oprah Magazine, House & Garden, Tin House, Ploughshares*, and *Bookforum*, among others. She has taught at Columbia, Bread Loaf, Bennington, New York University, and currently teaches in the MFA Program at The New School. She is editing an anthology on suburbia to be published by Vintage and is at work on a novel.

LIONEL SHRIVER has published seven novels, including *We Need to Talk About Kevin*, which won Britain's Orange Prize for Women Writers in 2005. Her work has appeared in the *New York Times*, the *Wall Street Journal*, the *Economist*, and the *Philadelphia Inquirer*. She divides her time between London and New York City and is currently at work on a romance novel.

LAUREN SLATER is the author of *Opening Skinner's Box: Great Psychological Experiments of the Twentieth Century, Prozac Diary, Welcome to My Country, Lying: A Metaphorical Memoir,* and *Love Works Like This: Travels Through a Pregnant Year,* among other books. Her work has appeared in the *New York Times Magazine, Elle, Harper's*, and many other publications.

LARRY SMITH has been the articles editor of *Men's Journal*, executive editor of *Yahoo! Internet Life*, senior editor at *ESPN* magazine, a founding editor of *P.O.V.*, and editor in chief of its sister publication, *Egg*, as well as an editor of *Might* magazine. While living in San Francisco, he was managing editor of the wire/syndication service AlterNet, and currently serves on the board of its umbrella organization, the Independent Media Institute. His work has appeared in the *New York Times*, *Wired, Popular Science, Salon, Men's Health, Best Life, SPIN, Vibe, Marie Claire, Teen People,* the *Columbia Journalism Review*, the *Utne Reader, Redbook*, the *Los Angeles Times*, and other publications.

LUISITA LÓPEZ TORREGROSA, an editor at the *New York Times*, is the author of the memoir *The Noise of Infinite Longing*. Her articles, essays, and reviews have appeared in the *New York Times, Vanity Fair, Condé Nast Traveler*, and *Vogue*. She is also a contributor to the anthology *The Literary Insomniac.* She lives in New York City.

CARY TENNIS has written *Salon*'s advice column, "Since You Asked," since 2001. He also plays music, writes fiction and poetry, and lives in San Francisco with his wife and two standard poodles.

REBECCA TRAISTER is a staff writer for *Salon*. She was formerly a reporter at the *New York Observer*. Her work has appeared in the *New York Times, Elle, Vogue, GQ*, and *New York* magazine. She lives in Brooklyn.

ALISA VALDES-RODRIGUEZ is the best-selling author of *The Dirty Girls Social Club, Playing With Boys, Make Him Look Good,* and the young adult novel *Haters*. She is an award-winning former staff writer for the *Boston Globe* and *Los Angeles Times*. A graduate of Columbia's journalism school, she is also a professional saxophonist and aerobics instructor. She and her husband and son divide their time between Miami and New Mexico.

ACKNOWLEDGMENTS

I FEEL PRIVILEGED TO WORK AT *SALON* with so many people whom I respect and adore. For their editorial advice and moral support throughout this project, I am grateful to all my colleagues, especially Joan Walsh, David Talbot, Ruth Henrich, Gary Kamiya, Scott Rosenberg, Max Garrone, Andrew Leonard, Kevin Berger, Kerry Lauerman, Amy Reiter, Andrew O'Hehir, Hillary Frey, Sarah Karnasiewicz, and Geraldine Sealey. I especially want to thank my colleague, partner in crime, and friend, Rebecca Traister, who makes coming to work each day so gratifying and fun.

To this book's talented contributors, I thank you for your honesty, your patience, and your willingness to bare all. My special appreciation goes to Anne Lamott for sharing her stories of motherhood with me, and for keeping me giggling at my keyboard with her hilarious e-mails.

I am fortunate to have had a series of mentors who have also become dear friends. For their advice, humor, and careful reading, I am grateful to Joan Walsh, Camille Peri, Laurie Abraham, Lisa Chase, and Dani Shapiro. My dear friends Elizabeth Kairys, Dawn Mackeen, Priscilla Yamin, Amy Goldwasser, Armin Harris, and Ali Penn patiently listened as I obsessed about this project for two years, and cheered me on until the very end. And thanks to Ilana Marcus and Carrie Ansell for twenty-eight years of friendship and guidance.

Sloan Harris at ICM believed in this book from the beginning and shepherded it through with care. Thanks also to Katherine Cluverius at ICM for her patience and help during every stage. Jeanette Perez at HarperCollins always had quick answers to my many questions and was a pleasure to work with. Erin Cox was a tireless advocate and a true believer in this book. And the fact that my excellent editor, Alison Callahan, gave birth, as did I, during the course of working on this book made our discussions about it more timely, personal, and vastly entertaining.

Without the unconditional support of my family, this book would not have been possible. My deepest appreciation goes to my parents, Joan Leibovich and Miguel Leibovich, and to Ted, Betty, Mark, Meri, Abby, and Marvin.

Finally, I give thanks to my husband, Larry Kanter, and to my son, Carlos Kanter: the two best decisions I ever made.